TERPENES AND TERPENOIDS

Edited by **Shagufta Perveen**
and **Areej Al-Taweel**

Terpenes and Terpenoids
http://dx.doi.org/10.5772/intechopen.71175
Edited by Shagufta Perveen and Areej Al-Taweel

Contributors

Ahmed Hussein, Alejandro Miguel Katzin, Rodrigo A.C Sussman, Emila A Kimura, Heloisa Berti Gabriel, Adriana Alejandra Marin Rodriguez, Ignasi Bonfill Verdaguer, Gabriela Carolina Fernandes Leite, Salud Perez, Roberto José Serrano Vega, Nimsi Campos Xolalpa, Angel Josabad Alonso Castro, Cuauhtémoc Pérez González, Julia Pérez Ramos, Shashikumar Paknikar, Kamlesh Pai Fondekar, Fozia Farhat, Arneeb Tariq, Annam Zikrea, Riffat Nasim Fatima, Shagufta Perveen, Areej Al-Taweel

First published in London, United Kingdom, 2018 by IntechOpen
IntechOpen is the global imprint of INTECHOPEN LIMITED, registered in England and Wales, registration number: 11086078, The Shard, 25th floor, 32 London Bridge Street
London, SE19SG – United Kingdom
Printed in Croatia

British Library Cataloguing-in-Publication Data
A catalogue record for this book is available from the British Library

Additional hard copies can be obtained from orders@intechopen.com

Terpenes and Terpenoids, Edited by Shagufta Perveen and Areej Al-Taweel
p. cm.
Print ISBN 978-1-78984-776-5
Online ISBN 978-1-78984-777-2

For EU product safety concerns:
IN TECH d.o.o., Prolaz Marije Krucifikse Kozulić 3, 51000 Rijeka, Croatia,
info@intechopen.com or visit our website at intechopen.com.

Meet the editors

Editor Dr. Shagufta Perveen is an Associate Professor at King Saud University in Saudi Arabia. She obtained her Ph.D. from the H.E.J. Research Institute of Chemistry, University of Karachi, Pakistan in 2009, and joined as an Assistant Professor at King Saud University then was promoted as an Associate Professor in 2014. She began her research on natural product isolation and structure elucidation using different modern techniques. She has published more than sixty original papers and three chapters in different ISI listed journals. She has collaborated with many researchers in Japan, United Kingdom, United Stated, Pakistan and Italy. She is the editor and reviewer of many journals. Currently, her main line of research is related to bio-guided isolation of secondary metabolites from plants, marine and plants endophytes and fungus.

Editor Prof. Dr. Areej Mohammad Al-Taweel obtained her Ph.D. in the field of Pharmacognosy from King Saud University in 2007. She then joined the same university as an Assistant Professor and was then promoted as an Associate Professor in 2013. She continues her research on phytochemical investigation on different Saudi Arabian medicinal plants and has isolated many bioactive natural products and published more than 40 papers in ISI ranking journals.

Contents

Preface

Terpenes belong to the diverse class of chemical constituents isolated from materials found in nature (plants, fungi, insects, marine organisms, plant pathogens, animals and endophytes). Theses metabolites have simple to complex structures derived from Isopentyl diphosphate (IPP), dimethyl allyl diphosphate (DMAPP), mevalonate and deoxyxylulose biosynthetic pathways. Terpenes play a very important role in human health and have significant biological activities (anticancer, antimicrobial, anti-inflammatory, antioxidant, anti-allergic, skin permeation enhancer, anti-diabetic, immunomodulatory, anti-insecticidal). According to new research, terpenes cineole (a spicy eucalyptus-derived flavoring oil) are ready to be directly converted to biofuel as soon as they are produced. This book provides an overview and highlights recent research in the phytochemical and biological understanding of terpenes and terpenoid and explains the most essential functions of these kinds of secondary metabolites isolated from natural sources.

Dr. Shagufta Perveen (MACS-USA, MRSC-UK, MIUPAC-USA)
Department of Pharmacognosy
College of Pharmacy
King Saud University
Riyadh, Saudi Arabia

Prof. Dr. Areej Al-Taweel
Department of Pharmacognosy
College of Pharmacy
King Saud University
Riyadh, Saudi Arabia

Chapter 1

Introductory Chapter: Terpenes and Terpenoids

Shagufta Perveen

Additional information is available at the end of the chapter

http://dx.doi.org/10.5772/intechopen.79683

1. Terpenes and terpenoids

Natural products are the compounds which isolate from different natural sources such as plants, animals, microbes, insects, plant pathogens, and endophytes and marine. These are known as secondary metabolites since they are formed due to the enzymatic resections of primary metabolites (amino acids, sugars, vitamins, etc.). Terpenes belong to the biggest

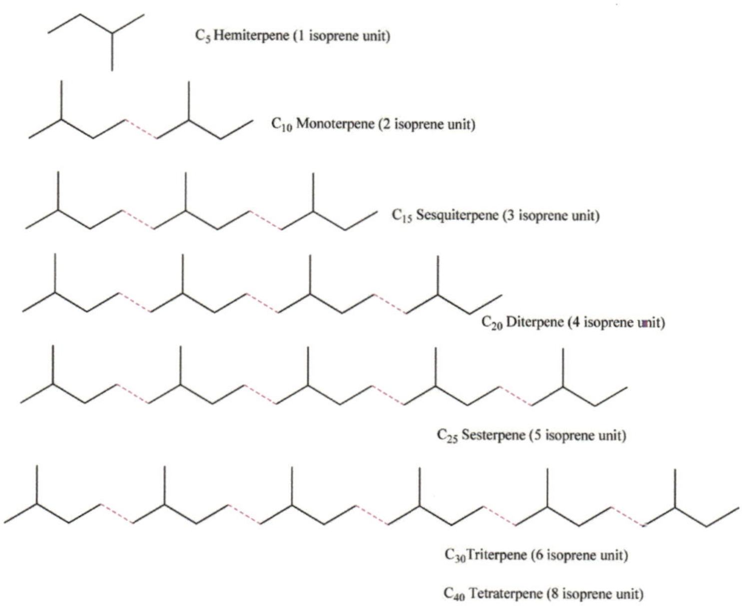

Figure 1. Classification of terpenes.

class of secondary metabolites and basically consist of five carbon isoprene units which are assembled to each other (many isoprene units) by thousands of ways. Terpenes are simple hydrocarbons, while terpenoids are modified class of terpenes with different functional groups and oxidized methyl group moved or removed at various positions. Terpenoids are divided into monoterpenes, sesquiterpenes, diterpenes, sesterpenes, and triterpenes depending on its carbon units (**Figure 1**). Most of the terpenoids with the variation in their structures are biologically active and are used worldwide for the treatment of many diseases. Many terpenoids inhibited different human cancer cells and are used as anticancer drugs such as Taxol and its derivatives. Many flavorings and nice fragrances are consisting on terpenes because of its nice aroma. Terpenes and its derivatives are used as antimalarial drugs such as artemisinin and related compounds. Meanwhile, terpenoids play a diverse role in the field of foods, drugs, cosmetics, hormones, vitamins, and so on. This chapter provides introduction and information on the bioactive terpenes isolated currently from different natural sources.

2. Monoterpenes

Monoterpenes consist of 10 carbon atoms with two isoprene units and molecular formula $C_{10}H_{16}$. These are naturally present in the essential and fixed oils of plants and related sources. Monoterpenes are structurally divided into the acyclic, monocyclic, and bicyclic type of

Names	Plant source	Activity	Ref.
9-OH-isoegomaketone [(2*E*)-1-(3-furanyl)-4-OH-4-Me-2-penten-1-one	Leaves of *Perilla frutescens var. crispa*	It exhibited inhibitory activity on nitric oxide (NO) production in lipopolysaccharide (LPS)-activated RAW264.7 cells with an IC_{50} value of 14.4 μM. Compounds in the SC-CO_2 extracts of the radiation mutant cultivar and the original plant were quantified by high-performance liquid chromatography with diode array detection.	[2]

Table 1. Source and biological activities of some monoterpenes.

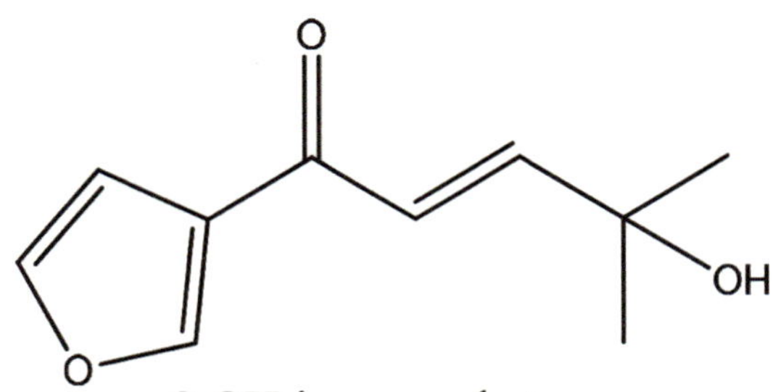

9-OH-isoegomaketone
[(2*E*)-1-(3-furanyl)-4-OH-4-Me-2-penten-1-one

Figure 2. Structure of monoterpene.

compound. The compounds belong to this class usually have strong aroma and odor and are used in many pharmaceutical companies. Mixture of different monoterpene-based oils is used as fragrances for making perfumes and in other cosmetics. Most of the monoterpenes are active biologically with strong antibacterial activities. Several studies have shown *in vitro* and *in vivo* antitumor activity of many essential oils obtained from plants. The antitumor activity of essential oils of many species has been related to the presence of monoterpenes in their composition [1]. Herein, we are discussing some of the recently published active monoterpenes (**Table 1**, **Figure 2**).

3. Sesquiterpenes

Sesquiterpenes are the class of secondary metabolites consisting of three isoprene units ($C_{15}H_{24}$) and found in linear, cyclic, bicyclic, and tricyclic forms. Sesquiterpenes are also found in the form of lactone ring (**Table 2**). Many of the latex in latex-producing plants contain sesquiterpene, and these are potent antimicrobial and anti-insecticidal agent. Artemisinin, a sesquiterpene lactone, one of the most active compounds in *Artemisia annua* shoots and roots (**Figure 3**).

Names	**Plant source**	**Activity**	**Ref.**
Arvestolides H and I	*Artemisia vestita*	H and I showed inhibitory effects on nitric oxide production in BV-2 cells induced by lipopolysaccharide with IC_{50} values of 43.2 and 39.9 μM, respectively.	[3]
Drimenin	Canelo tree *Drimys winteri*	Potency for drimenin at the hα4β2 AChR (0.97 μM) is several folds higher than that for other clinically used antidepressants using the same method. It could be used as a molecular scaffold for the development of more potent inhibitors with higher selectivity for the hα4β2 AChR.	[4]
Artefreynic acid B, C, and G	*Artemisia freyniana*	B, C, and G exhibited inhibitory effects against LPS-stimulated nitric oxide (NO) production in RAW 264.7 macrophage cells with IC_{50} values of 10.8, 12.6, and 11.7 μM, respectively.	[5]
Chrysanthemulide A	*Chrysanthemum indicum*	Mechanistic study revealed that the potential anti-inflammatory activity of A appears to be mediated via suppression of an LPS-induced NF-κB pathway and downregulation of MAPK activation.	[6]
14-*O*-Acetylinsulicolide A, 6β,9α-dihydroxy-14-p-nitrobenzoylcinnamolide, insulicolide A	Marine-derived *Aspergillus ochraceus* fungus	These compounds were evaluated for their cytotoxicities against three renal carcinoma cell lines, ACHN, OS-RC-2, and 786-O cells, and it displayed activities with IC_{50} values of 0.89–8.2 μM. Further studies indicated that it arrested the cell cycle at the G0/G1 phase at a concentration of 1 μM and induced late apoptosis at a concentration of 2 μM after a 72 h treatment of 786-O cells.	[7]

Names	Plant source	Activity	Ref.
Santhemoidin A	*Tarchonanthus camphoratus* and *Schkuhria pinnata*	A was the most active compound found in this study, with IC_{50} values of 0.10 μM against *Trypanosoma brucei rhodesiense* trypomastigotes and selectivity indices of 20.5, respectively.	[8]

Table 2. Source and biological activities of some sesquiterpenes.

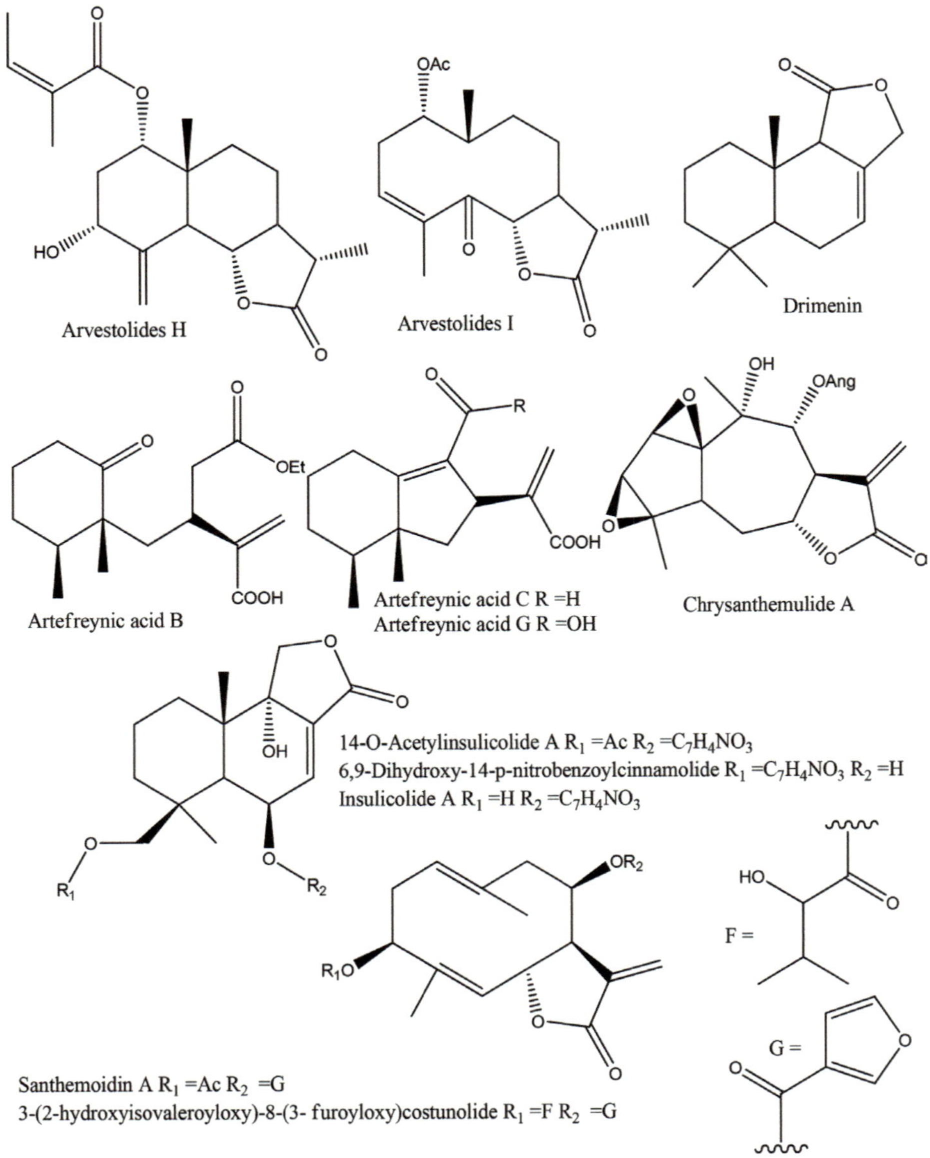

Figure 3. Structures of sesquiterpenes.

4. Diterpenes

Diterpenoids belong to a versatile class of chemical constituents found in different natural sources having $C_{20}H_{32}$ molecular formula and four isoprene units (**Figure 4**). This class of compounds showed significant biological activities including anti-inflammatory, antimicrobial,

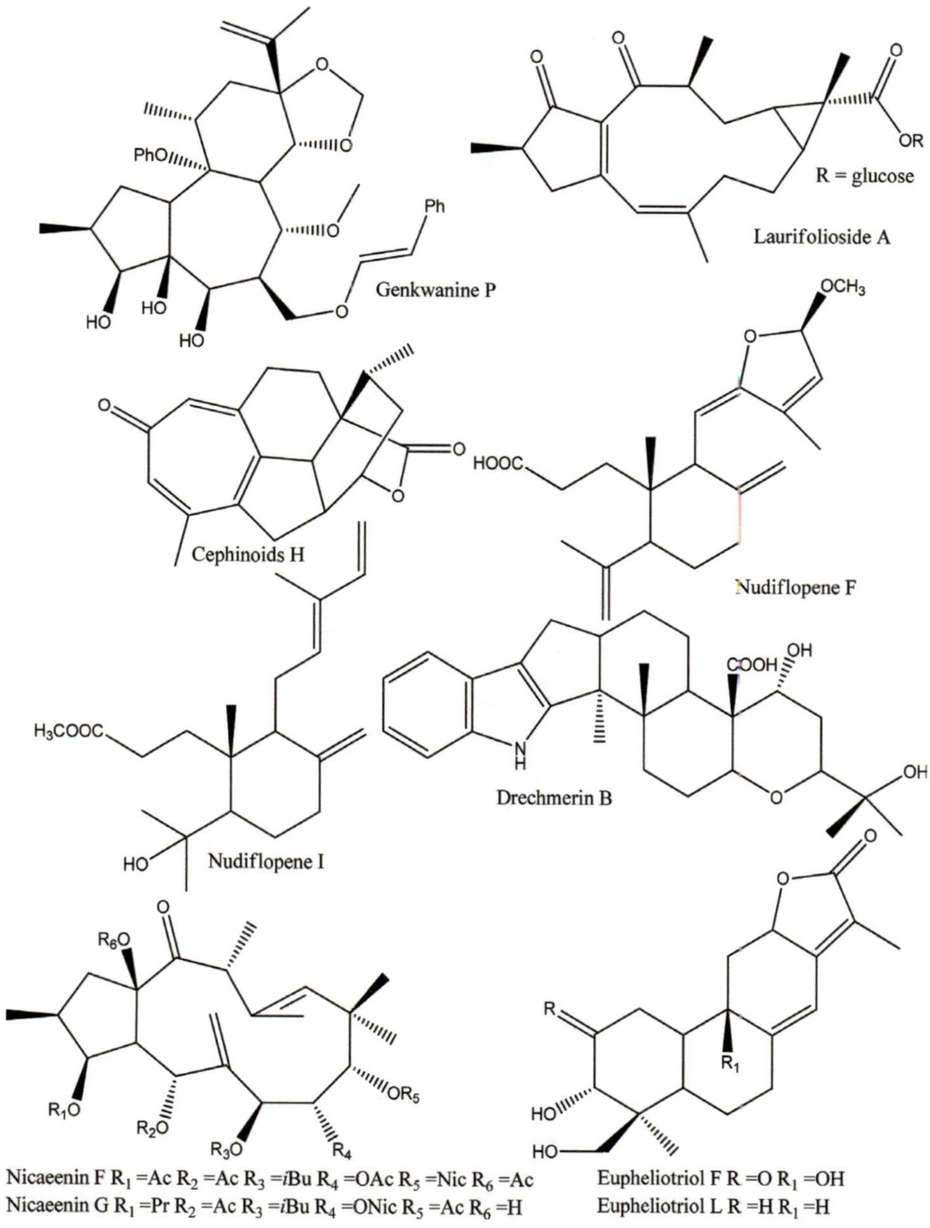

Figure 4. Structure of diterpenes.

Names	Plant source	Activity	Ref.
Genkwanine P and laurifolioside A	Buds of *Wikstroemia chamaedaphne*	Compounds exhibited potential antihepatitis B virus activities with IC_{50} 46.5 and 88.3 mg/mL against HBsAg.	[9]
Cephinoids H	*Cephalotaxus fortunei* var. alpina and *C. lanceolata*	H demonstrated an inhibition of 49.0% by administration to zebra fish at a dose of 60.0 ng/mL, compared to cisplatin (DDP, 22.4%) at 15.0 μg/mL. It might affect the NF-κB signaling pathway rather than binding to microtubules. Additionally, it showed almost equal anti-inflammatory activities compared to the positive control, MG132.	[10]
Nudiflopene F and I	Leaves of *Callicarpa nudiflora*	F and I have strong interactions with the iNOS protein by targeting residues of the active cavities of iNOS n BV-2 cells (IC_{50} 28.1 and 23.3).	[11]
Drechmerin B	Endophytic fungus Drechmeria sp.	B displayed antimicrobial activity against *C. albicans* with an MIC value of 12.5 μg/mL.	[12]
Nicaeenin F	Latex of *Euphorbia nicaeensis*	F showed significant potential to inhibit P-glycoprotein (P-gp) activity in two MDR cancer cells (NCI-H460/R and DLD1-TxR).	[13]
Nicaeenin G	Latex of *E. nicaeensis*	G showed significant potential to inhibit P-glycoprotein (P-gp) activity in two MDR cancer cells (NCI-H460/R and DLD1-TxR). G also significantly stronger chemosensitized NCI-H460/R cells to DOX compared to Dexverapamil due to prolonged effect of P-gp inhibition that persisted for 72 h.	[13]
Eupheliotriol F and L	*Euphorbia helioscopia*	F and L exhibited significant cytotoxicity against MCF-7 and PANC-1 cell lines.	[14]

Table 3. Source and biological activities of some diterpenes.

anticancer, and antifungal activities. Some of the diterpenes also have cardiovascular activity, such as grayanotoxin, forskolin, eleganolone, marrubenol, and 14-deoxyandrographolide. Kaurane and pimarane-type diterpenes are also biologically active metabolites isolated from the roots and leaves of different plants (**Table 3**).

5. Sesterpenes

Sesterpenes consist of 25 carbon atoms with 5 isoprene units and molecular formula $C_{25}H_{40}$ (**Figure 5**). These are naturally present in the fungus, marine organism, insects, sponges, lichens, and protective waxes of insects. These types of compounds are biologically active having anti-inflammatory, anticancer, antimicrobial, and antifungal activities (**Table 4**).

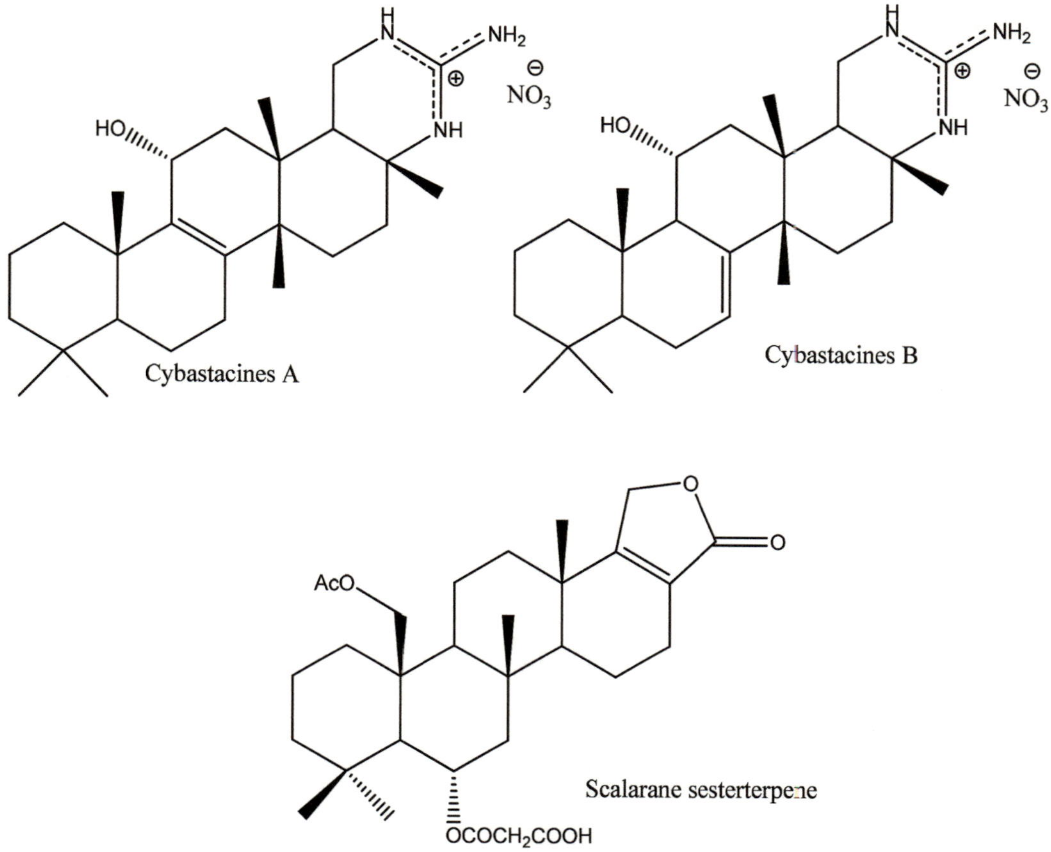

Figure 5. Structures of sesterpenes.

Names	Plant source	Activity	Ref.
Cybastacines A and B	*Nostoc* sp. Cyanobacterium	A and B showed moderate in vitro antibiotic activities. Sesterterpenes are rare among microbial secondary metabolites, with only one report of a previous alkaloid—sesterterpene found in cyanobacteria. This discovery represents a significant addition to the novel chemical structures active against resistant bacterial strains.	[15]
Scalarane sesterterpenes	Mushroom species, *Pleurotus ostreatus* and *Scleroderma areolatum*	This compound exhibited moderate micromolar activity against *P. falciparum* 3D7 and *T. cruzi* Tulahuen C4 parasites. It showed <50% inhibition at 25 μM, when incubated with the tumoral liver cell line, HepG2 (HB-8065) for 72 h.	[16]

Table 4. Source and biological activities of some sesterpenes.

6. Triterpenes

A major class of secondary metabolites are known as triterpenes and it usually contains 30 carbon atoms consisting of 6 isoprene units (**Figure 6**). It is derived from the squalene biosynthetic pathway. Triterpenes have many methyl groups and it can be oxidized into alcohols, aldehydes, and carboxylic acids, which make it complex and differentiate it biologically. Triterpenes have many active sites for the glycosylation which converts it into another big class of compounds, namely, saponins (triterpene glycoside). Herein, we are discussing some recently published bioactive triterpenes (**Table 5**).

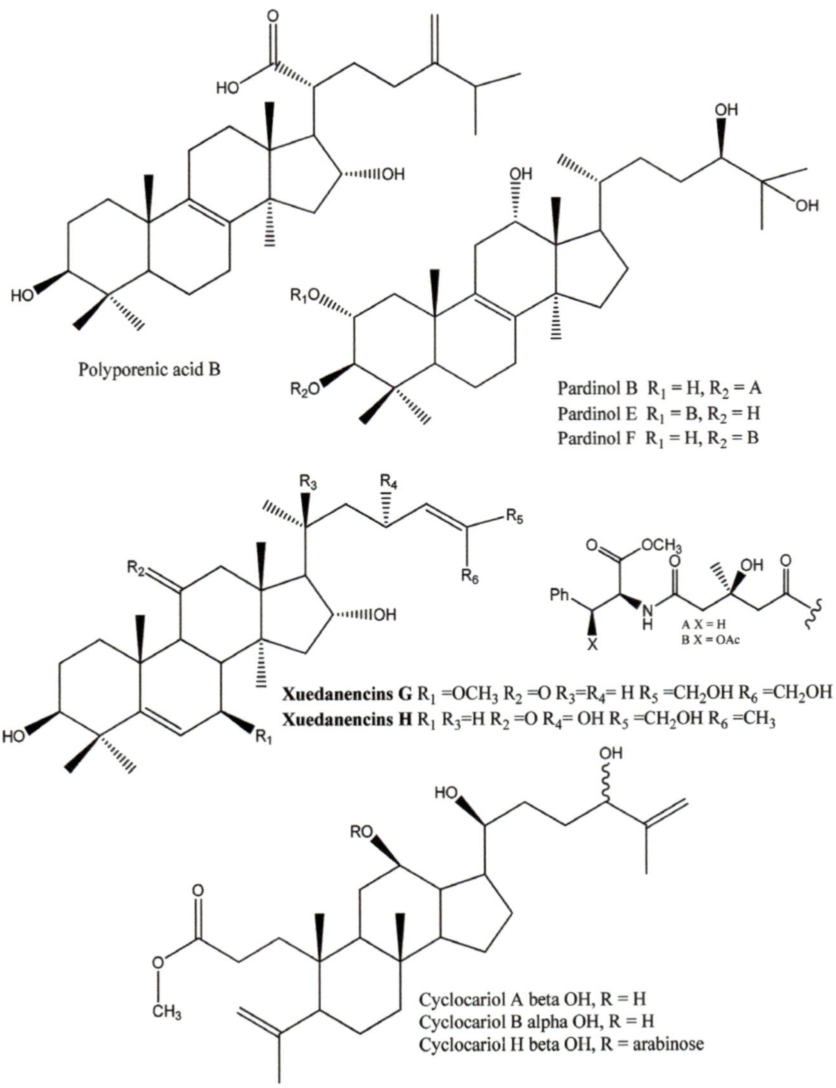

Figure 6. Structure of triterpenes.

Names	Plant source	Activity	Ref.
Polyporenic acid B	Fruiting bodies of *Fomitopsis palustris*	It showed strong cytotoxicity against the HCT116, A549, and HepG2 cell lines with IC_{50} values of 8.4, 12.1, and 12.2 μM, respectively.	[17]
Pardinol B	*Tricholoma pardinum*	Compound showed strong cytotoxicity against HL-60 SMMC-7721 A-549 MCF-7 SW480, 8.3, 15.0, 14.4, 12.7, 15.0 μM, respectively.	[18]
Pardinol E	T. pardinum	E exhibited strong cytotoxicity against HL-60 SMMC-7721 A-549 MCF-7 SW480, 9.8, 11.7, 9.8 11.9, 15.6, μM, respectively	[18]
Pardinol F	*T. pardinum*	F showed strong cytotoxicity against HL-60 SMMC-7721 A-549 MCF-7 SW480, 11.2, 15.6, 12.6, 10.5, 14.1 μM, respectively.	[18]
Xuedanencins G and H	Tubers of *Hemsleya penxianensis*	G and H were evaluated for cytotoxic activity against the Hela human cancer cell line and compounds showed significant cytotoxicity with IC_{50} value at 1.82 and 2.45 μM, respectively.	[19]
Cyclocariols A, B, and H	Leaves of *Cyclocarya paliurus*	A, B, and H were tested against human colon tumor (HCT-116) cell lines, exhibited good activities with IC_{50} values of 6.53, 4.94, and 6.48 μM, respectively.	[20]

Table 5. Source and biological activities of some triterpenes.

7. Meroterpenes

Meroterpenes are the secondary metabolites with partial terpenoid skeleton. Meroterpenoids were partially derived from mevalonic acid pathways and widely derived from animals, plants,

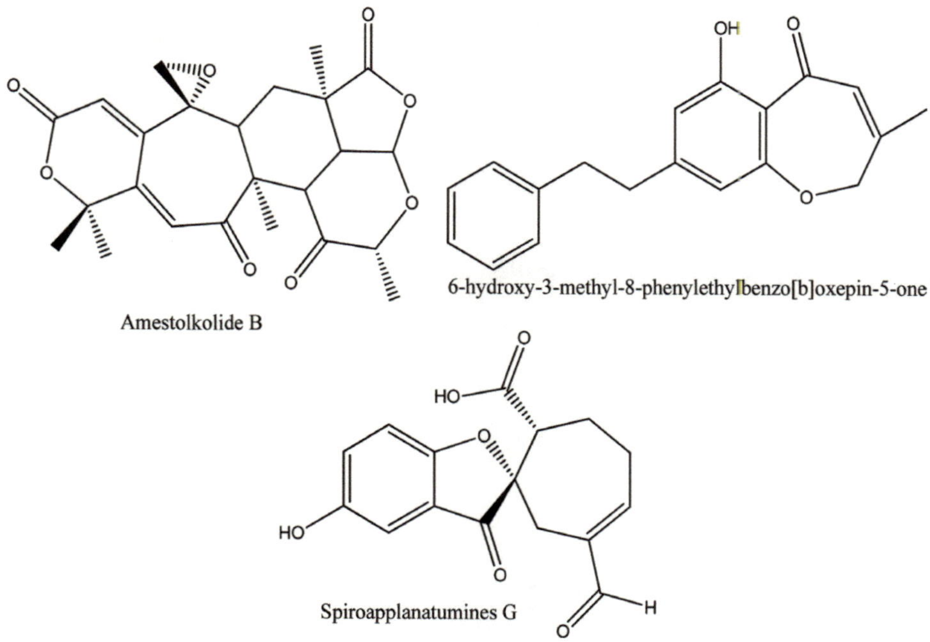

Figure 7. Structures of meroterpenes.

Names	Plant source	Activity	Ref.
Amestolkolide B	Mangrove endophytic fungus *Talaromyces amestolkiae* YX1	B showed strong anti-inflammatory activity in vitro by inhibiting nitric oxide (NO) production in lipopolysaccharide activated in RAW264.7 cells with IC_{50} value of 1.6 ± 0.1 mM.	[22]
6-OH-3-Me-8-phenylethylbenzo[b]oxepin-5-one	Liverwort *Radula sumatrana*	This compound showed activity against the human cancer cell lines MCF-7, PC-3, and SMMC-7721, with IC_{50} values of 3.86, 6.60, and 3.58 μM, respectively, and induced MCF-7 cell death through a mitochondria-mediated apoptosis pathway.	[23]
Spiroapplanatumines G	*Ganoderma applanatum*	Biological evaluation of compound 7 inhibited JAK3 kinase with IC_{50} values of 7.0 μM	[24]

Table 6. Source and biological activities of some meroterpenes.

bacteria, and fungi [21] (**Figure 7**). Meroterpene biosynthesis expands the diversity available to isoprenoid pathways alone and allows for the assembly of natural products with highly unique structural attributes. Organisms belonging to the fungal kingdom have become proficient at exploiting this broad chemical synthesis platform for complex metabolite production. Herein, we are discussing some of the recently published bioactive meroterpenes (**Table 6**).

Acknowledgements

This research project was supported by a grant from the "Research Center of the Female Scientific and Medical Colleges," Deanship of Scientific Research, King Saud University.

Author details

Shagufta Perveen

Address all correspondence to: shagufta792000@yahoo.com

Department of Pharmacognosy, College of Pharmacy, King Saud University, Riyadh, Saudi Arabia

References

[1] Marianna VS, Aline LX, Tamires CL, Damiao PS. Antitumor activity of monoterpenes found in essential oils. Scientific World Journal. 2014;**953451**:1-35. DOI: 10.1155/2014/953451

[2] Bomi N, Yangkang S, Hyo-Young K, Jin-Baek K, Chang HJ, Ah-Reum H. A new monoterpene from the leaves of a radiation mutant cultivar of *Perilla frutescens* var. *crispa* with inhibitory activity on LPS-induced NO production. Molecules. 2017;**22**:1471-1472. DOI: 10.3390/molecules22091471

[3] Shuai-Hua T, Chen Z, Ke-Wu Z, Ming-Bo Z, Yong J, Peng-Fei T. Sesquiterpenoids from *Artemisia vestita*. Phytochemistry. 2018;**147**:194-202. DOI: 10.1016/j.phytochem.2018.01.00e4

[4] Hugo RA, Dominik F, Bernd S, Matthias H, Cristian P, Marcelo OO. Drimane sesquiterpenoids noncompetitively inhibit human α4β2 nicotinic acetylcholine receptors with higher potency compared to human α3β4 and α7 subtypes. Journal of Natural Products. 2018;**81**:811-817. DOI: 10.1021/acs.jnatprod.7b00893

[5] Chen Z, Ran W, Xiao-Li M, Ke-Wu Z, Yang X, Pu-Ming Z, Ming-Bo Z, Yong J, Guo-Qing L, Peng-Fei T. Nitric oxide inhibitory sesquiterpenoids and its dimers from *Artemisia freyniana*. Journal of Natural Products. 2018;**81**:866-878

[6] Gui-Min X, Xiao-Qing L, Chen C, Kang C, Xiao-Bing W, Yu-Cheng G, Jian-Guang L, Ling-Yi K. Highly oxidized guaianolide sesquiterpenoids with potential antiinflammatory activity from *Chrysanthemum indicum*. Journal of Natural Products. 2018;**81**:378-386. DOI: 10.1021/acs.jnatprod.7b00867

[7] Yanhui T, Bin Y, Xiuping L, Xiaowei L, Xiaoyan P, Lan T, Yonghong L, Xiaojuan L, Xuefeng Z. Nitrobenzoyl sesquiterpenoids with cytotoxic activities from a marine-derived *Aspergillus ochraceus* fungus. Journal of Natural Products. 2018;**81**:92-97. DOI: 0.1021/acs.jnatprod.7b00698

[8] Njogu MK, Josphat CM, Marcel K, Reto B, Thomas JS. Antiprotozoal sesquiterpene lactones and other constituents from *Tarchonanthus camphoratus* and *Schkuhria pinnata*. Journal of Natural Products. 2018;**81**:124-130. DOI: 10.1021/acs.jnatprod.7b00747

[9] Shi-Fei L, Ying-Ying J, Zhi-Qiang Z, Jian-Bin C, Jie J, Xun-Long S, Li-Wei Z. Diterpenes from buds of *Wikstroemia chamaedaphne* showing antihepatitis B virus activities. Phytochemistry. 2018;**151**:17-25. DOI: 10.1016/j.phytochem.2018.01.021

[10] Ling N, Xiu-Hong Z, Xiao-Jiao C, Bing-Jie Z, Mei-Fen B, Xiang-Hai C. Bioactive nor-diterpenoids from *Cephalotaxus fortunei* var. *alpina* and *C. lanceolate*. Phytochemistry. 2018;**151**:50-60. DOI: 10.1016/j.phytochem.2018.04.007

[11] Xiaocong S, Feng L, Xueyuan Y, Jinghan W, Bangjian D, Chunfeng X, Da-Qing J, Jie Z, Dongho L, Yasushi O, Jing X, Yuanqiang G. Seco-labdane diterpenoids from the leaves of *Callicarpa nudiflora* showing nitric oxide inhibitory activity. Phytochemistry. 2018;**149**:31-41. DOI: 10.1016/j.phytochem.2018.02.001

[12] Jian-Chao Z, Ya-Li W, Tian-Yuan Z, Zhong-Jian C, Tian-Mei Y, Ying-Ying W, Cheng-Peng S, Xiao-Chi M, Yi-Xuan Z. Indole diterpenoids from the endophytic fungus *Drechmeria* sp. as natural antimicrobial agents. Phytochemistry. 2018;**48**:21-28. DOI: 10.1016/j.phyto chem.2018.01.010

[13] Gordana K, Milka J, Nina MT, Milica P, Tijana S, Ivana SA, Vele VT. Jatrophane diterpenoids with multidrug-resistance modulating activity from the latex of *Euphorbia nicaeensis*. Phytochemistry. 2018;**148**:104-112. DOI: 10.1016/j.phytochem.2018.01.016

[14] Wan-Ping W, Kun J, Ping Z, Kai-Kai S, Shi-Jin Q, Xiao-Ping Y, Chang-Heng T. Highly oxygenated and structurally diverse diterpenoids from *Euphorbia helioscopia*. Phytochemistry. 2018;**145**:93-102. DOI: 10.1016/j.phytochem.2017.10.012

[15] Alfredo HC, Víctor TP, Santiago MC, Diego FRV, Antera MQ, Montserrat OD, Angel RS. Cybastacines A and B: Antibiotic sesterterpenes from a *Nostoc* sp. cyanobacterium. Journal of Natural Products. 2018;**81**:410-413. DOI: 10.1021/acs.jnatprod.7b00638

[16] Frederick A, Ignacio P, Theresa A, Guiomar P-M, Elizabeth D, Jesús M, Thomas M, Luis R, Dolores G, Olga G, Francisca V, Christian A, Fernando R. Antiprotozoan sesterterpenes and triterpenes isolated from two Ghanaian mushrooms. Fitoterapia. 2018;**127**:341-348. DOI: 10.1016/j.fitote.2018.03.016

[17] Jinzhi Z, Yun Y, Mengyao Y, Ke Y, Xia L, Huayi Q, Guolin Z, Yinggang L. Lanostane-type C31 triterpenoid derivatives from the fruiting bodies of cultivated *Fomitopsis palustris*. Phytochemistry. 2018;**152**:10-21. DOI: 10.1016/j.phytochem.2018.04.012

[18] Shuai-Bing Z, Zheng-Hui L, Marc S, He-Ping C, Ying H, Xiao-Qing GTF, Ji-Kai L. Lanostane triterpenoids from *Tricholoma pardinum* with NO production inhibitory and cytotoxic activities. Phytochemistry. 2018;**152**:105-112. DOI: 10.1016/j.phytochem.2018.05.002

[19] Nailiang Z, Zhonghao S, Meigeng H, Yedan L, Dawei Z, Haifeng W, Yu T, Pengfei L, Junshan Y, Guoxu M, Xudong X. Cucurbitane-type triterpenes from the tubers of *Hemsleya penxianensis* and their bioactive activity. Phytochemistry. 2018;**147**:49-56. DOI: 10.1016/j.phytochem.2017.12.014

[20] Yan-jiao C, Liang N, Jialong F, Jianping Z, Nusrat H, Yu-qing J, Hanwen Y, Bin L, Bin L, Iqbal C, Ikhlas K, Wei W. Seco-dammarane triterpenoids from the leaves of *Cyclocarya paliurus*. Phytochemistry. 2018;**145**:85-92. DOI: 10.1016/j.phytochem.2017.10.013

[21] Senhua C, Meng D, Weiyang L, Xishan H, Zhaoming L, Yongjun L, Hongju L, Zhigang S. Anti-inflammatory meroterpenoids from the mangrove endophytic fungus *Talaromyces amestolkiae* YX1. Phytochemistry. 2018;**146**:8-15. DOI: 10.1016/j.phytochem.2017.11.011

[22] Senhua C, Meng D, Weiyang L, Xishan H, Zhaoming L, Yongjun L, Hongju L, Zhigang S. Anti-inflammatory meroterpenoids from the mangrove endophytic fungus *Talaromyces amestolkiae* YX. Phytochemistry. 2018;**146**:8-15. DOI: 10.1016/j.phytochem.2017.11.011

[23] Xiao W, Lin L, Rongxiu Z, Jiaozhen Z, Jinchuan Z, Hongxiang L. Bibenzyl-based meroterpenoid enantiomers from the Chinese liverwort *Radula sumatrana*. Journal of Natural Products. 2017;**80**:3143-3150. DOI: 10.1021/acs.jnatprod.7b00394

[24] Qi L, Xiao-Yi W, Jing Y, Jin-Feng L, Rui L, Zheng-Chao T, Yong-Xian C. Spiro meroterpenoids from *Ganoderma applanatum*. Journal of Natural Products. 2017;**80**:61-70. DOI: 10.1021/acs.jnatprod.6b00431

Chapter 2

Chemistry of South African Lamiaceae: Structures and Biological Activity of Terpenoids

Ahmed A. Hussein

Additional information is available at the end of the chapter

http://dx.doi.org/10.5772/intechopen.77399

Abstract

South Africa flora is one of the most important mega floras with high endemic species percentage. Lamiaceae is an important family in South Africa with ±308 species in 41 genera and contains many important plants (~23%) traditionally used for treatment of different human diseases. The chemical profile of Lamiaceae is very rich in terpenoids in general and more specifically diterpenes. Genera like *Leonotis* and *Plectranthus* are well studied, while on the other hand, genus like *Stachys* (~41 species, ~50% endemic) didn't receive any attention. Different classes of diterpenes were identified and some of them demonstrating important biological activities.

Keywords: South African flora, Lamiaceae, *Leonotis*, *Plectranthus*, chemical constituents, terpenoids

This work is dedicated to Prof. Benjamin Rodriguez (Instituto de Quimica Organic General, CSIC, Spain) for his contributions in the field of natural products and specially in the chemistry of Lamiaceae family.

1. Introduction

The Green economy concept has been driven as an urgent need for addressing global challenges in vital fields like energy, environment, and health. Green economy is expected to play a very important role in changing the way that society manages the interaction of the environmental and economic domains. Consequently, a new paradigm has been established and shifted toward green economy or green growth. Natural products represent one of the most important elements required to build safe and effective economy especially in health sector. South Africa (SA) is recognized as one of the most biodiverse country in the world with 20,456 indigenous vascular plant taxa recorded where 13,265 (65%) are endemic [1, 2].

The Lamiaceae (formerly Labiateae, mint family) is a cosmopolitan family with ~7136 species in 236 genera. Most species are shrubby or herbaceous and trees are extremely rare [3]. The Lamiaceae family has great economic value, as it contains several horticultural species, most of which are used as culinary herbs like salvia, rosemary, ocimum, mint, *Leonotis*, etc. Lamiaceae species are known to contain pharmacologically active terpenoids with a wide spectrum of bioactivity and expected to play more important roles in the process of drug discovery as well as cosmetic, food, and pesticides industries [4–6]. In the Sub-Saharan region, ~60 genera with ±980 species were reported [7]. SA considers as a diversity spot of Lamiaceae with ±308 species in 41 genera [8]. The species occur predominantly in the summer and/or winter rainfall areas. The habitats are different and vary to a great extent [9].

However, the South African flora is one of the most important mega floras for its unique diversity and endemism, it receives low attention in terms of bioprospecting, and the number of research paper every year dealing with chemical/biological profiling is still beyond the required level. This review serves as a background for the chemistry of all species belonging to the family Lamiaceae growing in SA and it covers publications till 2017. The articles information's abstracted from Sci-finder database [10] and includes all species growing in SA as well as other places. This chapter doesn't cover the essential oils and *Plectranthus barbatus*, which recently reviewed by others [11, 12].

2. Terpenoids of different genera of South African Lamiaceae

Different classes of secondary metabolites have been identified from Lamiaceae, the majority of the isolated compounds are terpenoids (~71%), and additionally other classes of compounds like flavonoids, α-pyrone derivatives, phenolic acids, and alkaloids were reported. Mono-, sesqui-, and tri-terpenoids are relatively small in number (~15%) when compared to diterpenoids and it was reported that more than 100 of different diterpene skeletons were identified which indicate the high evolutionary index of Lamiaceae [13]. According to the literature, the genera *Leonotis* (known as wild dagga) and *Plectranthus* have received the highest attention where 70 (*Leonotis*) and 94 (*Plectranthus*) compounds were identified so far, the majority of the isolated compounds are labdane diterpenes. In this chapter, the different genera have been listed alphabetically and the trivial names have been retained in the cases where they were given by authors and/or chemical abstracts.

2.1. *Aeollanthus* genus

Aeollanthus genus represented by 43 species globally and 7 in SA. From *A. buchnerianus*, an abietanediterpene, [(rel)-14α-acetoxyabiet-7-en-18-oicacid] (**1**) [14], 3β-acetoxy-7,15-isopimaradiene

(**2**), 3*β*-acetoxy-7,15-isopimaradien-19-ol (**3**) and 19-acetoxy-7,15-isopimaradien-3*β*-ol (**4**), 7,15-isopimaradien-19-ol (**5**, akhdarenol) and 7,15-isopimaradien-3*β*,19-diol (**6**, virescenol), a mixture of 19-isobutyryloxy- and 19-butyryloxy-8*β*-hydroxy-15-isopimarene (**7**), and a 3:1 mixture of 5-stigmasten-3*β*-ol and *β*-sitosterol were isolated from the aerial parts of *A. rydingianus*. 5 and 6 showed activity against *S. aureus* and *Enterococcus hirae* [15].

2.2. *Ballota* genus

Ballota is represented by one species in SA *vizB africana*. Hispanolone (**8**) was isolated from the aerial parts [16].

2.3. *Cedronella* genus

Cedronella genus is represented by only one species in SA *viz C. canariensis*. The phytochemical studies of the aerial parts resulted in isolation of a dimer of *d*-pinocarvone (**9**), cedronellone (**10**), and ursolic acid (**11**) [17].

2.4. *Clerodendrum* genus

Seven species were recorded in SA and clerodendrumic acid (**12**) was isolated from *C. glabrum* var. *glabrum* and showed weak antifungal, antibacterial, and cytotoxic activities [18].

2.5. *Hoslundia* genus

Hoslundia genus is represented by one species in SA *vizH. opposite*. The phytochemical studies of the aerial parts yielded an interesting and rare pyrano and furanoflavonoid derivatives in addition to euscaphic and (**13**) ursolic acid (**11**) [19, 20]; four abietane-type esters, 3-*O*-cinnamoylhosloppone (**14**), 3-*O*-benzoylhosloppone (**15**), 3-*O*-benzoylhosloquinone (**16**), and 3-*O*-benzoylhinokiol (**17**); 13 was found to exhibit MIC of 50 μg/mL against *M. tuberculosis*, while 14 inhibits the growth of the MDR strain K_1 of *Plasmodium falciparum* in vitro with an IC_{50}-value of 0.4 μg/mL [21].

2.6. *Hyptis* genus

Three species were recorded in SA. The triterpenes 3*α*,19*α*-dihydroxyurs-12-en-28-oic acid (**18**) and 3*β*-acetoxyoleanan-13*β*,28-olide (**19**), Me betulinate (**20**), oleanolic acid/acetate (**21/22**), and ursolic (**11**) and maslinic acids (**23**) were isolated from *H. mutabilis* [22].

From *H. spicigera*, seven labdane diterpenes; 19-acetoxy-2*α*,7*α*,15-trihydroxylabda-8(17),(13*Z*)-diene (**24**); 15,19-diacetoxy-2*α*,7*α*-dihydroxylabda-8(17),(13*Z*)-diene (**25**); 7*α*,15,19-triacetoxy-2*α*-hydroxylabda-8(17),(13*Z*)-diene (**26**); 19-acetoxy-2*α*,7*α*-dihydroxylabda-8(17),(13*Z*)-dien-15-al (**27**); 19-acetoxy-7*α*,15-dihydroxylabda-8(17),(13*Z*)-dien-2-one (**28**); 2*α*,7*α*,15,19-tetrahydroxy-ent-labda-8(17), (13*Z*)-diene (**29**); and 19-acetoxy-2*R*,7*R*-dihydroxylabda-14,15-dinorlabd-8(17)-en-13-one (**30**) were isolated from the aerial parts [23].

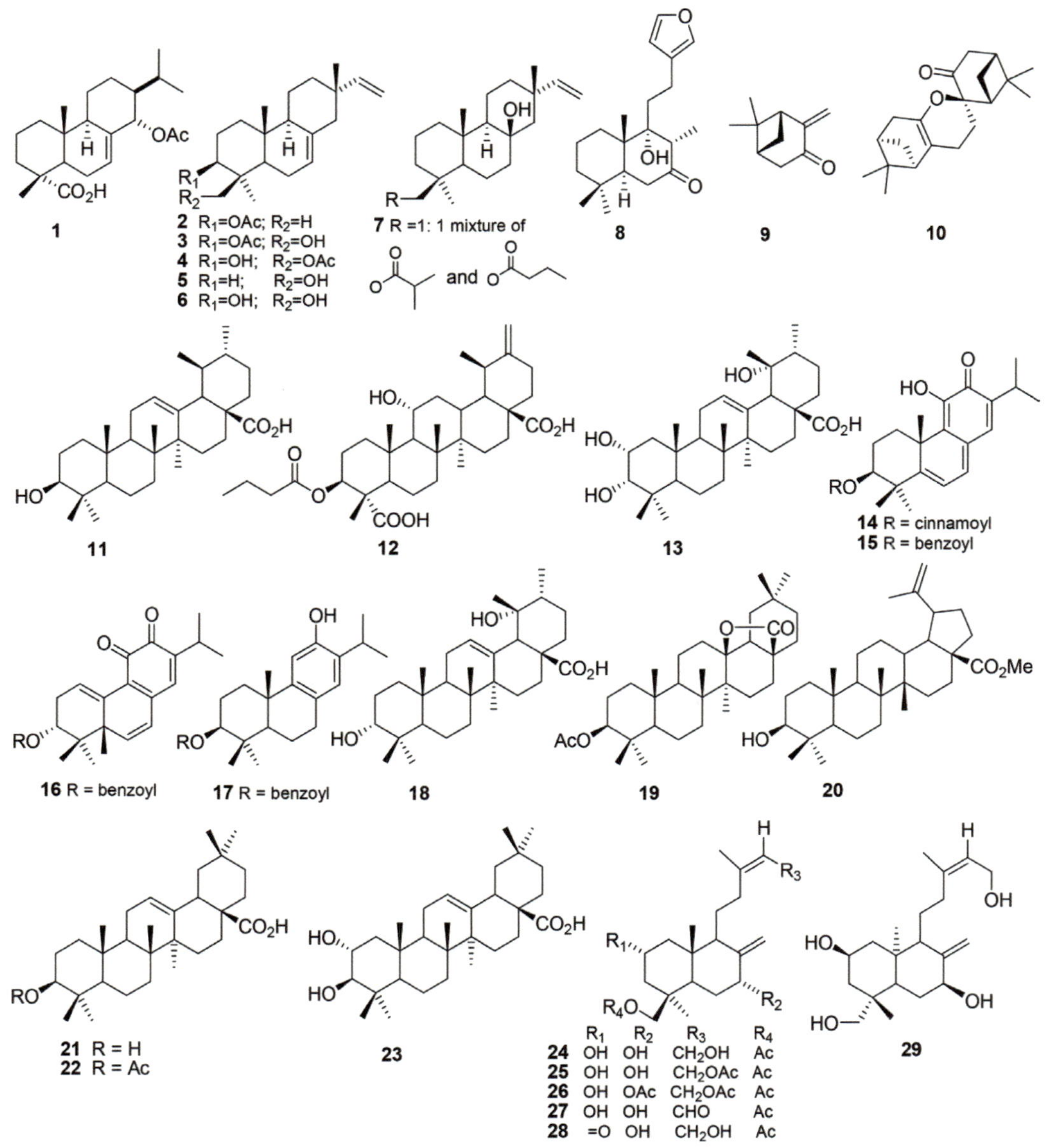

2.7. *Leonotis* genus

Seven species were recorded in SA and two of them were extensively studied. Traditionally, this genus is used to substitute hemp and called as wild dagga; however, there is no much scientific biological evidences supporting such claim. The chemistry was started in early 60s of the last century by South African researchers. Many labdane diterpenes have been isolated. The chemistry of the genus was covered previously by a review published by Piozzi et al.[24].

2.7.1. *Leonotis leonurus*

The chemistry of *Leonotis* was commenced in 1962 and some compounds were identified; marrubiin (**31**) compounds, X (**32**) and Y (**33**), the stereoisomers of premarrubiin (**34**) and (**35**) (the *C*-13 epimeric forms of premarrubiin). Leonurun (**36**) has been isolated and the relative stereochemistry was determined using single-crystal X-ray diffraction analysis [24, 25]. After two years, labdane (13*S*)-9α,13α-epoxylabda-6β(19),15(14)-dioldilactone (**37**) was isolated, this compound caused significant changes in blood pressure of anesthetized normotensive rats, and also was found to exhibit a negative chronotropic effect [26].

The organic extract of *L. leonurus* showed 99% growth inhibition against *M. tuberculosis* at 1.0 mg/mL, subsequent phytochemical studies resulted in the identification of three labdane-type diterpenoids: 9,13:15,16-diepoxy-6,16-labdanediol (**38**), 6-acetoxy-9,13-epoxy-15-methoxy-labdan-16,15-olide (**39**), and 9,13-epoxy-6-hydroxylabdan-16,15-olide (**40**). None of the isolated compounds were active against *M. tuberculosis* [27].

Recently, Fang et al. [28] identified leonurenones A–C (**41–43**), in addition to 9,13:15,16-diepoxy-6,16-labdanediol (**38**) and nepetifolin (**44**). The leonurenones contain an uncommon α,β-unsaturated enone moiety in ring B. Compound **38** was isolated as epimeric form, (at C-16, ratio 3:1). Compound **41** was isolated from aqueous extract of the leaves and the authors proposed the possible formation of **43** as an artefact *via* oxidation and lactonization of the more polar intermediate (**41**) during the isolation process. The total aqueous extract, at concentration of 1.0 g/mL, showed an 81% inhibition in a binding assay at the GABAA site. Compounds **41** and **43** did not show activity (<50% inhibition) in this assay [28].

In the following year, Wu and co-workers (2013) were successful to isolate and identify eleven labdanoides, *viz* leoleorins D–J (**41–43, 45–48**) and 16-epi-leoleorin F (**49**), leoleorin A [corresponding to compound Y (**33**)], leoleorin B (**50**) (anhydro derivative of compound Y), and leoleorin C [9,13-epoxy-6-hydroxylabdan-15,16-olide (**40**)]. The absolute configurations of leoleorin A (**33**) and D (**41**) were established by X-ray crystallographic analyses. It is important to indicate that new compounds "leoleorins G-I", which were isolated in this study, were reported in the previous work under the names of leonurenones A–C (**41–43**) (^{13}C data showed exchange positions C_{12} and C_{14} for leonurenones C/leoleorin H between the two references) [29].

From *L. leonurus*' flowers, an acyclic diterpene ester, 1,2,3-trihydroxy-3,7,11,15-tetramethyl-hexadecan-1-yl-palmitate (**51**), along with geniposidic acid (**52**) were isolated, the compounds exhibited neither cytotoxicity on mammalian kidney fibroblasts (Vero cells) nor antimicrobial activities [30].

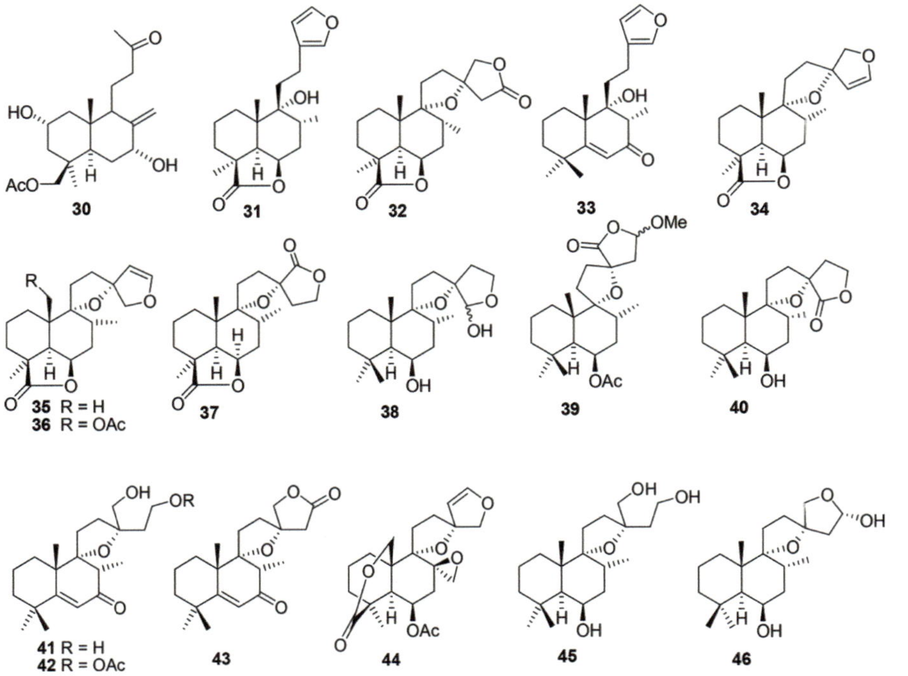

2.7.2. Leonotis nepetaefolia

The chemistry of *L. nepetaefolia* started almost simultaneously with *L. leonurus*. Leonotin (**53**), nepetaefuran (**54**), nepetaefuranol (**55**), nepetaefolin (**44**) methoxynepetaefolin (**56**), nepetaefolinol (**57**) and leonotinin (**58**) the dilactone (8*β*,17,9,13-diepoxylabdane-16,15,19,6*β*-diolactone, **59**) were characterized [31–36].

From the species collected from India, nepetaefolinol (**57**), dehydrated nepetaefolinol (**60**) and isomeric tetrol (**61**) (15,16-epoxy-labda-13(16),14-diene-6*β*,9,17,19-tetrol: the reduction product of leonotinin) were identified [37]. Leonitinic acid (**62**) with free C-17 carboxyl group was also isolated [38].

From a commercially material, originally collected from Peru, five inseparable epimeric mixtures of bis-spirolabdane diterpenoids, resulted from biosynthetic epimerization of three different structures around C-13 and C-15, have been isolated and identified as leonepetaefolin A (**63**) and its epimeric isomer 15-epi-leonepetaefolin A (**64**) (ratio 1:1), leonepetaefolin B(**65**)/15-epi-leonepetaefolin B (**66**) (2:3), leonepetaefolin C(**67**)/15-epi-leonepetaefolin C (**68**) (1,1), leonepetaefolin D (**69**)/15-epi-leonepetaefolin D (**70**) (7,10), leonepetaefolin E (**71**)/15-epi-leonepetaefolin E (**72**) (2,3) [39]. Additionally, methoxynepataefolin (**56**), nepetaefolin (**44**), nepetaefuran (**54**), dubiin (**73**), 19 chlroro derivative of nepetaefolin (**74**), leonotinin (**58**), leonotin (**53**), and LS-1 (**75**) were isolated. The absolute configuration of the epimeric mixture **63** and **64** was determined by X-ray crystallographic analysis [39].

http://dx.doi.org/10.5772/intechopen.77399

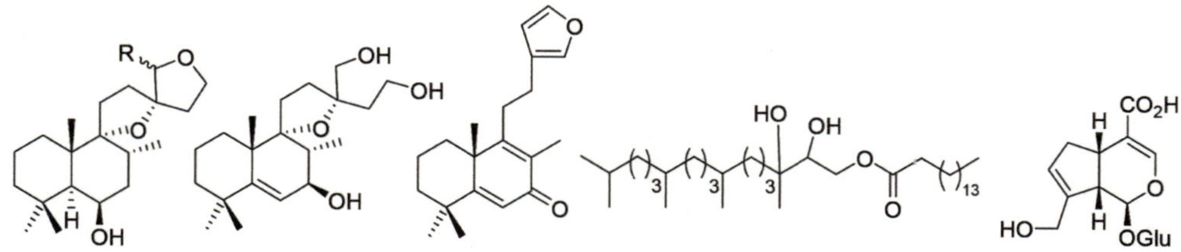

47 R = βOH
49 R = αOH
48 **50** **51** **52**

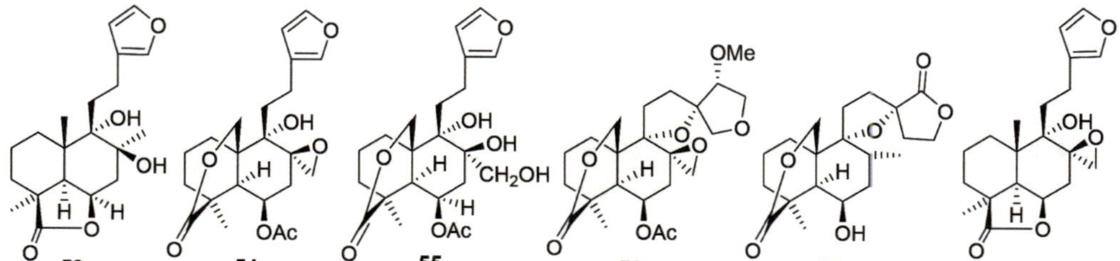

53 **54** **55** **56** **57** **58**

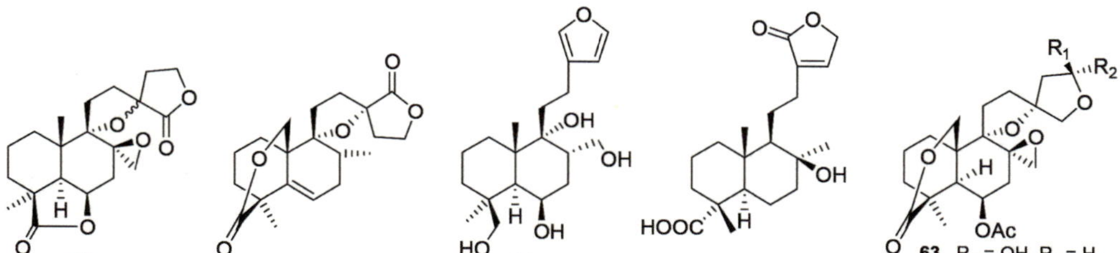

59 **60** **61** **62**
63 R_1 = OH R_2 = H
64 R_1 = H R_2 = OH

The isolated compounds were evaluated for their binding activities to a panel of CNS G-protein-coupled receptors including adrenergic, dopaminergic, histaminic, muscarinic, opioid, and serotonergic receptors and neurotransmitter transporters and showed no interesting activity.[39]. From the material collected from Japan, five iridoid glycosides: 10-*O*-(*trans*-3,4-dimethoxycinnamoyl) geniposidic acid (**76**), 10-*O*-(*p*-hydroxybenzoyl) geniposidic acid (**77**), geniposidic acid (**52**), mussaenoside (**78**), and ixoside (**79**) were isolated [40].

2.7.3. *Leonotis ocymifolia*

L. ocymifolia was studied under different synonyms *viz; L. dubia* (*L. ocymifolia*, var. *ocymifolia*), *L. leonitis; L. leonitis* var. *hirtfolia* (*L. ocymifolia*, var. *ocymifolia*) and *L. dysophylla* Benth. (*L. ocymifolia* var. *raineriana*) and *L. ocymifolia* var. *raineriana* (Burm f) Iwarsson var. *raineriana* (Visiani) Iwarsson. The chemical studies resulted in the isolation of dubiin (**73**), 9*α*,13(*S*)-epoxy-8*β*-hydroxylabdane-6*β*,19;16,15-diolide (**80**), and leonitin (**81**). 20-acetoxy-9*α*,l3-dihydroxy-15(16)-epoxylabd-14-en-6*β*(19)-lactone (**82**) and 6*β*-acetoxy-9*α*,l3*α*-epoxylabda-20(19),16(15)-diol-dilactone (**83**) are from the leaves, in addition to compound X (**32**)[24, 41] Finally, nepetaefolin (**44**), leonotinin (**58**), and leonotin (**53**) were identified from the material collected from Pretoria (South Africa) [42].

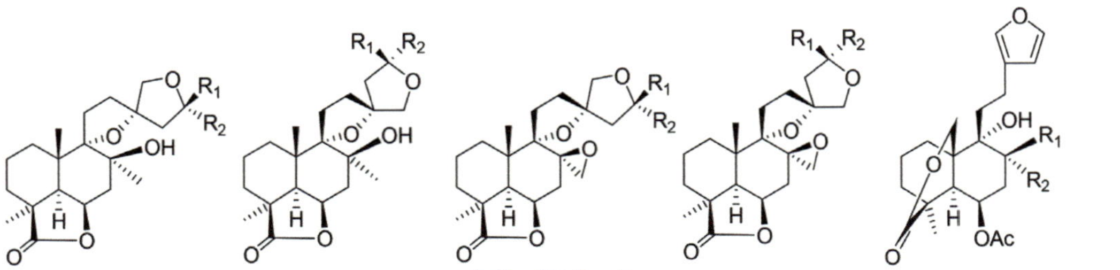

65 R_1= OH, R_2 = H
66 R_1= H, R_2 = OH
67 R_1= OH, R_2 = H
68 R_1 = H, R_2 = OH
69 R_1 = OH, R_2 = H
70 R_1 = H , R_2 = OH
71 R_1 = OH, R_2 = H
72 R_1 = H, R_2 = OH
73 R_1= H, R_2 = CH_3
74 R_1= OH, R_2 = CH_2Cl

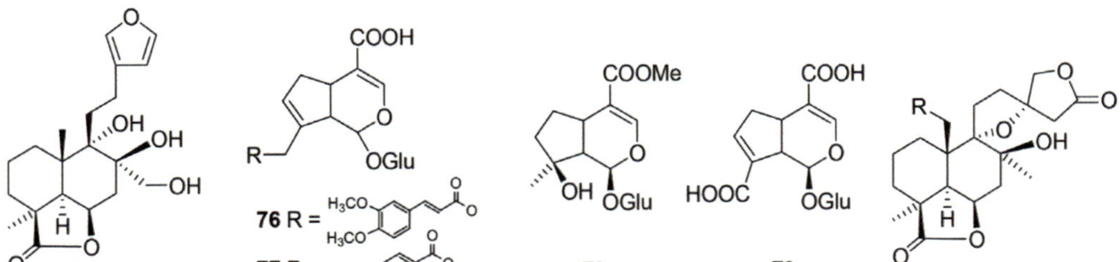

75
76 R =
77 R =
78
79
80 R = H
81 R = OAc

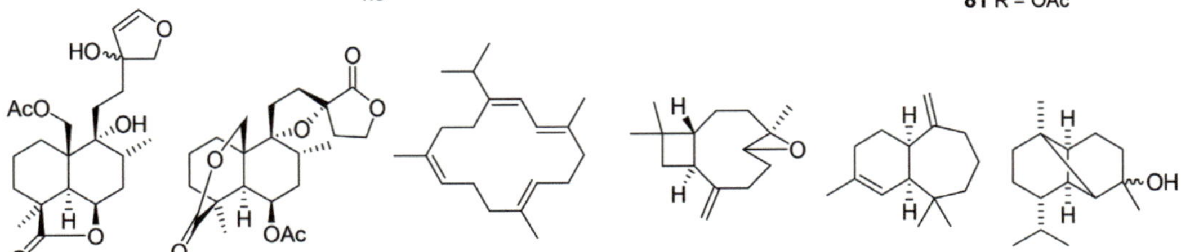

82 **83** **84** **85** **86** **87**

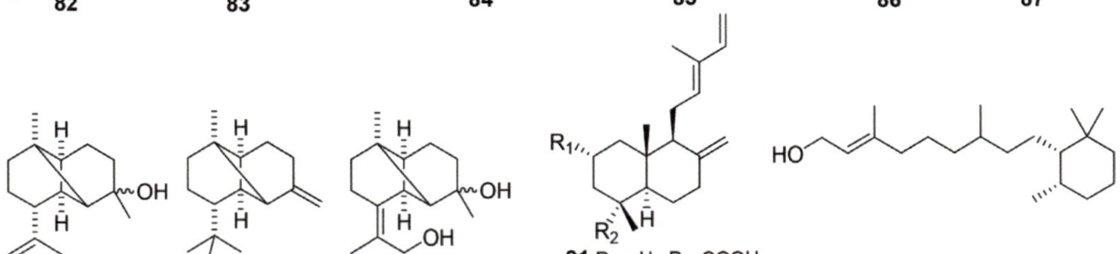

88 **89** **90**
91 R_1 = H; R_2=COOH
92 R_1 =OH; R_2=CH_2OH
93 R_1 =OH; R_2=COOH
94

2.8. *Neophyptis* genus

Neophyptis genus is represented by *N paniculata* in SA. Isoneocembrene-A (**84**), *β*-caryophyllene oxide(**85**), *α*-himachalene (**86**), the isolates showed weak to moderate antibacterial activity against five strains of *S. aureus* [43].

2.9. *Ocimum* genus

Ocimum genus comprises 65 aromatic species, distributed in tropical and subtropical regions worldwide. Species belonging to this genus are popularly used in Africa and Asia for treating diabetic symptoms. The genus is represented by 16 species in SA and the phytochemical study of *O. amercanium* afforded four compounds of the copane series (copan-3-ol (**87**), cop-11(12)-en-3-o1 (**88**), cop-3(15)-en-11-ol (**89**), and cop-l0(ll)-en-3,12-diol(**90**)) [44].

2.10. *Orthosiphon* genus

Orthosiphon genus comprises 40 species recorded from the old world: in tropical and sub-tropical regions including Southern Africa and Madagascar. Three species were found in SA. Three labdanoids (+)-*trans*-ozic acid (**91**), labda-8(17),12*E*,14-trien-2α,18-diol (**92**), and 2α-hydroxylabda-8(17),12*E*,14-trien-18-oic acid (**93**) have been isolated from an ethanol extract. Compound **93** exhibited activity against *M. tuberculosis*, while **92** showed cytotoxic activity against MCF-7 and decreased the production of all the pro-inflammatory cytokines. From the same source, pheophytin a, the acidic degradation product of chlorophyll a, was isolated and showed inhibition of HIV-1 protease [45, 46].

2.11. *Paltstoma* genus

Only one species was recorded in SA. From the ethyl acetate extract of *P. rotundifolium*, cas-sipourol (**94**), β-sitosterol, and α-amyrin were identified [47].

2.12. *Plectranthus* genus

About 300 species distributed in tropical and warm regions of the old World, 45 species recorded in SA, from which 19 species were studied for their chemical and/or biological constituents. The genus is characterized by the presence of orange glands that distributed in the aerial parts and contain highly oxygenated (and modified) abietane-type diterpenoids. Others, e.g., kaurane, labdane, phyllocladane as well as the rare skeleton halimane diterpenoids were described.

2.12.1. Plectranthus ambiguus

The plant afforded a series of tetracyclic phyllocladane-type (= 13β-kaurane) diter-penoids: (16*R*)-2α-senecioyloxy-3α-acetoxyphyllocladan-16,17-diol (**95**), (16*R*)-2α-senecioyloxy-3α,17-diacetoxy-16-hydroxyphyllocladane (**96**), (16*R*)-2α-isovaleroyloxy-3α-acetoxyphyllocladan-16,17-diol (**97**), (16*R*)-2α-isovaleroyloxy-3α,17-diacetoxy-16-hydroxyphyllocladane (**98**), (16*R*)-3α-acetoxyphyllocladan-16,17-diol (**99**), (16*R*)-2α-senecioyloxy-16,17-dihydroxyphyllocladan-3-one (**100**), and (16*R*)-2α,3α-diacetoxyphyllocladan-16,17-diol (**101**). The authors discriminated between phyllocladane and *ent*-kaurane tetracyclic skeletons after extensive spectroscopic inves-tigation as well as chemical transformations [48, 49].

2.12.2. Plectranthus amboinicus

Thymoquinone (**105**) was identified as an active nonpolar ingredient to suppress the expres-sion of lipopolysaccharide-induced tumor necrosis factor-alpha (TNF-α) [50]. The total extract showed cytotoxic activity against MCF-7, using HPLC-based metabolomics approach, and 7α-acetoxy-6β-hydroxyroyleanone (**102**) was identified as the main active constituent. Other minor compounds like coleon E (**103**) and royleanone (**104**) were also identified [51].

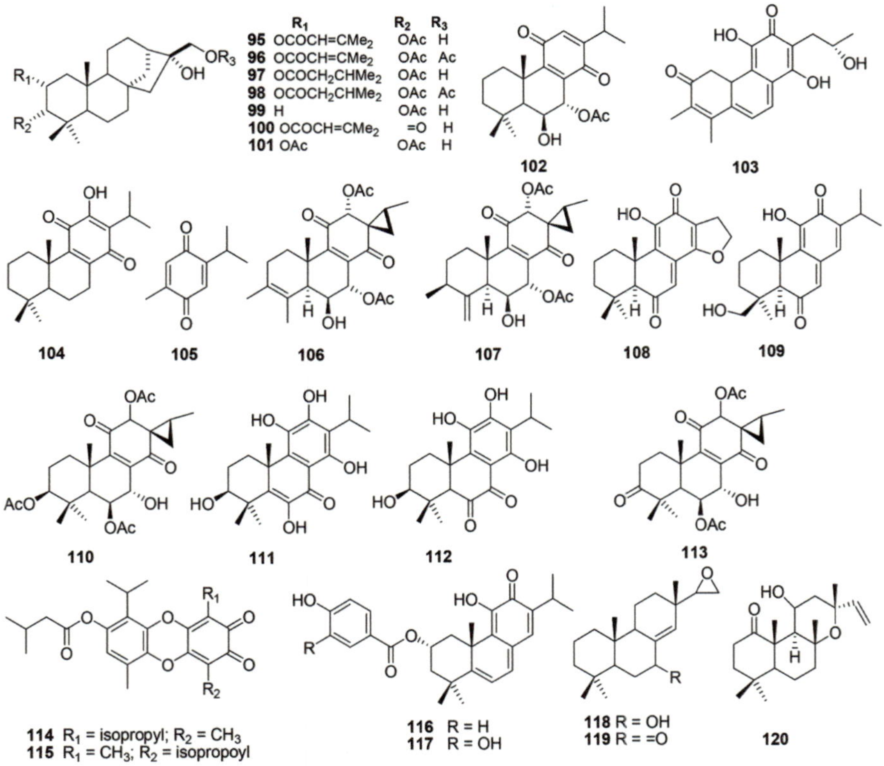

2.12.3. *Plectranthus caninus*

Plectranthus caninus afforded coleons M (**106**), N (**107**), P (**108**), Q(**109**), R (**110**), S (**111**), and T (**112**) and barbatusin (**113**) [52, 53].

2.12.4. *Plectranthus ecklonii*

Plectranthus ecklonii is traditionally used in South Africa for treating stomach aches, nausea, vomiting, and meningitis. Ecklonoquinone A (**114**) and B (**115**) and parviflorons D (**116**) and F (**117**) were isolated [54, 55]. Compound **117** showed potent activity against *Listeria monocytogenes* and *M. tuberculosis* and both **116** and **117** were found to be very toxic against vero cell lines. The potency of parvifloron D (**116**) was further confirmed and showed fast and potent apoptotic inducer in leukemia cells [56].

2.12.5. *Plectranthus ernstii*

Two pimaranes rel-15(ζ),16-epoxy-7α-hydroxypimar-8,14-ene (**118**): rel-15(ζ),16-epoxy-7-oxopimar-8,14-ene (**119**) and a labdane 1*R*,11*S*-dihydroxy-8*R*,13*R*-epoxylabd-14-ene (**120**) were isolated. The three compounds showed activity against *M. tuberculosis* and different strains of *S. aureus* [57].

2.12.6. *Plectranthus fruticosus*

Plectranthus fruticosus cultivated in Porugal afforded 4 labdanes, *ent*-labda-8(17),12*Z*,14-trien-2β-ol (**121**),*ent*-2α-acetoxylabda-8(17),12*Z*,14-trien-3β-ol (**122**), ent-3β-acetoxylabda-8(17),

12Z,14-trien-2α-ol (**123**),3β-acetoxylabda-8(17),12E,14-trien-2α-ol (**124**). 10 kauranes (*ent*-12β-acetoxy-15β,16β-epoxykauran-19-oic acid (**125**), *ent*-7β-hydroxy-15β,16β-epoxykauran-19-oic acid (**126**), *ent*-15β,16β-epoxykauran-19-oic acid (**127**), *ent*-15β,16β-epoxykauran-19-ol (**128**), *ent*-12β-acetoxy-15β-hydroxykaur-16-en-19-oic acid (**129**), *ent*-12β-acetoxy-7β-hydroxykaur-16-en-19-oic acid (**130**),methyl ent-12β-acetoxy-16-kauren-19-oate (**131**), *ent*-7β-hydroxykaur-15-en-19-oic acid (**132**), methyl *ent*-12β-acetoxy-7β-hydroxykaur-15-en-19-oate acid (**133**), *ent*-12β-acetoxy-17-oxokaur-15-en-19-oic acid (**134**), methyl *ent*-12β-acetoxy-15-kauren-19-oate (**135**), additionally, armendrance (**136**), caryophyllene α-oxide (**137**), ursolic/oleanolic acids (2,1 mixture) β-sitosterol, stigmasta-5,22*E*-dien-3β-ol, and β-amyrin. Some of the compounds showed moderate anti-*staphylococcus* activity [58, 59]. *P. fruticosus* growing in India showed abietane diterpene pattern and 7α-acetoxy-6β-hydroxyroyleanone (**102**), 6,7-dehydroroyleanone (**138**) and 7α,6β-dihydroxyroyleanone (**139**) were isolated [60].

2.12.7. *Plectranthus grandidentatus*

In addition to 14-hydroxytaxodione (**140**), coleons U (**141**) and V (**142**), a series of abietane dimers namely grandidone A (**143**), B(**145**), and D(**147**) and their epimers 7-epigrandidone A(**144**), B(**146**), and D (**148**) and grandidone C (**149**) [61] were identified. Also, royleanone (**103**), 6,7-dehydroroyleanone (**138**), horminone (**150**), 6β-hydroxyroyleanone (**151**), and 7α-acetoxy-6β-hydroxyroyleanone (**102**) together with a mixture of fatty acid esters of 7α-acyloxy-6β,12-dihydroxy-abieta-8,12-diene-11,14-dione (**152**), 7α,6β,-dihydroxyroyleanone (**139**), and 9α-(2-oxopropyl)abietane derivative(**156**) were isolated [62–67].

Fatty acid esters of 7α-acyloxy-6β-hydroxyroyleanone (**152**) showed moderate antibacterial activity [62]; coleon U exhibited potent cytotoxicity against a panel of human cancer cell lines [63, 65] also showed potent inhibition of mouse splenocyte proliferation induced by ConA or LPS mitogens [64]. Coleons U **141** is considered as a promising compound and deserves further evaluation as an anti-cancer drug [68]. Coleon U (**141**), 7α-acetoxy-6β-hydroxyroyleanone (**102**), and horminone (**150**) showed activity against methicillin-resistant *S. aureus* (MRSA) and vancomycin-resistant *Enterococcus faecalis* (VRE). Recently, the biological activity of **102** was reported and showed selective cytotoxicity against MCF-7. Other derivatives of the same compound showed potent cytotoxic [69, 70] and antimicrobial [66] activities.

2.12.8. *Plectranthus hereroensis*

Horminone (**150**), 16-acetoxy-7α,12-dihydroxy-8,12-abietadiene-11,14-dione (**153**) and 7α-12-dihydroxy-17(15→16)-abieta-8,12,16-triene-11,14-dione (**157**);3β-acetoxy-6β,7α-12-trihydroxy-17(15→16)18(4→3)bisabeo-abieta-4(19)8,12,16-triene-11,14-dione (**158**) were isolated [13, 66, 71], on the other hand, the structure of an aristolane sesquiterpene aldehyde (**159**) have been revised [72], all compounds showed moderate antimicrobialactivity [13, 66, 71, 72], while **158** showed antiviral activity [73].

2.12.9. *Plectranthus madagascariensis*

Plectranthus madagascariensis is used as a traditional medicine in Southern Africa. Three constituents were isolated and identified as 6β,7β-dihydroxyroyleanone (**154**), 7β-acetoxy-6β-hydroxyroyleanone (**155**), and coleon U (**141**). The compounds exhibited inhibitory activity on α-glucosidase, *S. aureus* and *Enterococcus faecalis* [74].

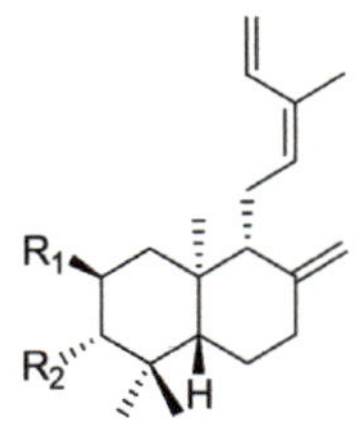

121 R_1 = OH, R_2 = H
122 R_1 = OAc, R_2 = OH
123 R_1 = OH, R_2 = OAc

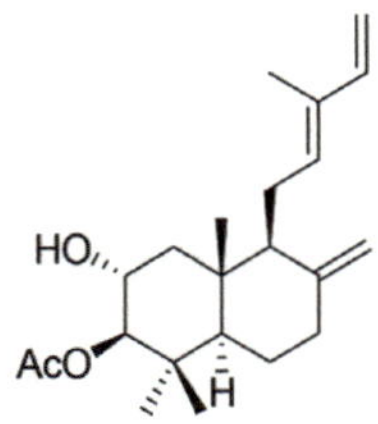

124

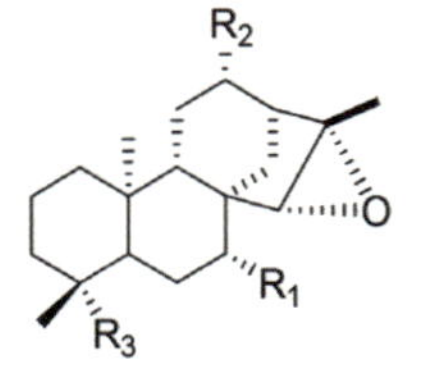

125 R_1 = H, R_2 = OAc, R_3 = CO_2H
126 R_1 = OH, R_2 = H, R_3 = CO_2H
127 R_1 = R_2 = H, R_3 = CO_2H
128 R_1 = R_2 = H, R_3 = CH_2OH

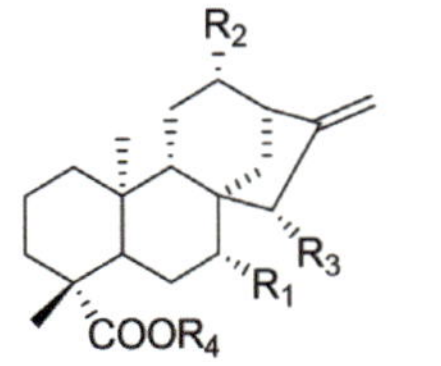

129 R_1 = R_4 = H, R_2 = Ac, R_3 = OH
130 R_1 = OH, R_2 = Ac, R_3 = R4 = H
131 R_1 = R_2 = R_3 = H, R4 = Me

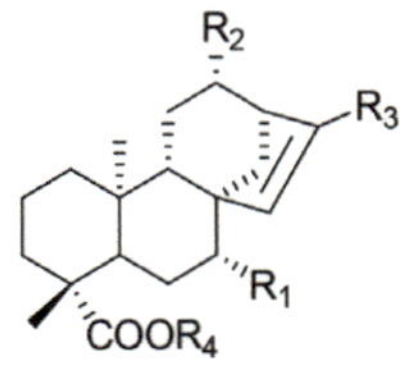

132 R_1 = OH, R_2 = R_4 = H, R_3 = Me
133 R_1 = OH, R_2 = OAc, R_3 = Me, R_4 = Me
134 R_1 = R_4 = H, R_2 = OAc, R_3 = CHO
135 R_1 = H, R_2 = OAc, R_3 = R_4 = Me

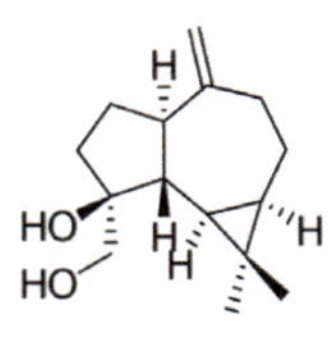

136

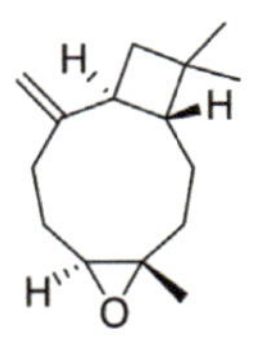

137

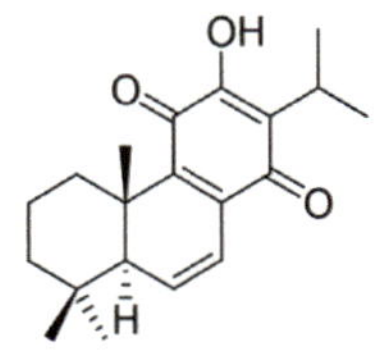

138

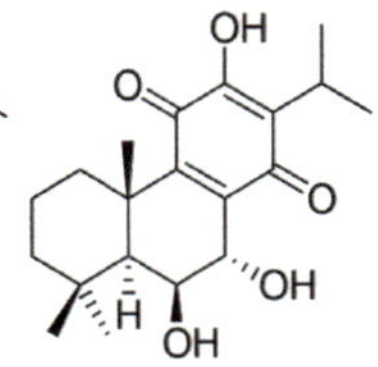

139

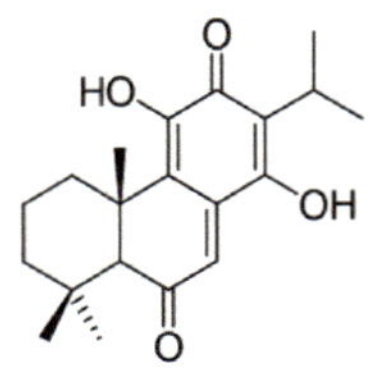

140

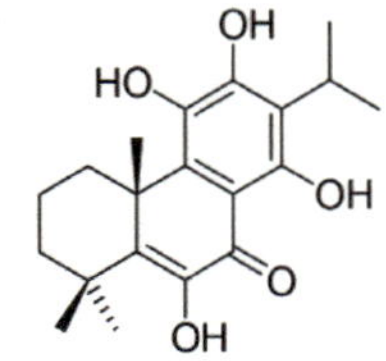

141

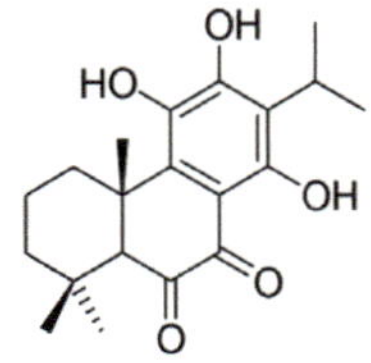

142

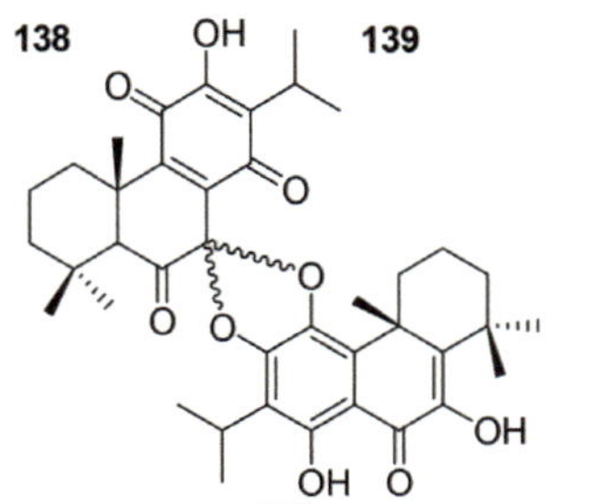

143 β-C(7)-O-C(11')
144 β-C(7)-O-C(12')

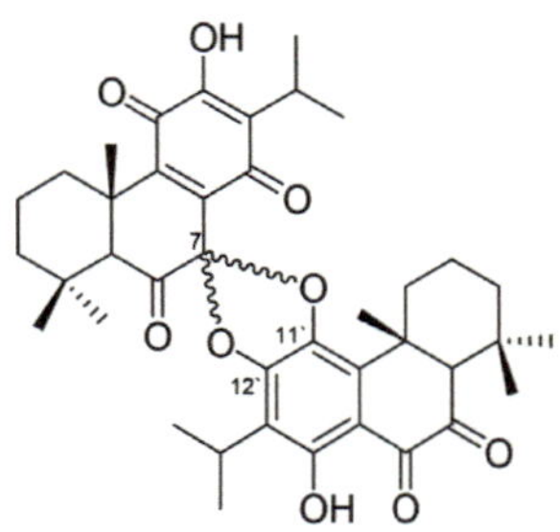

145 β-C(7)-O-C(11')
146 β-C(7)-O-C(12')

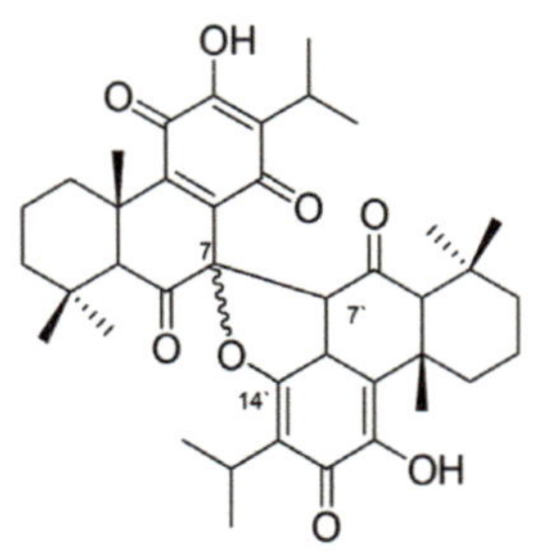

147 α-C(7)-O-C(14')
148 β-C(7)-O-C(14')

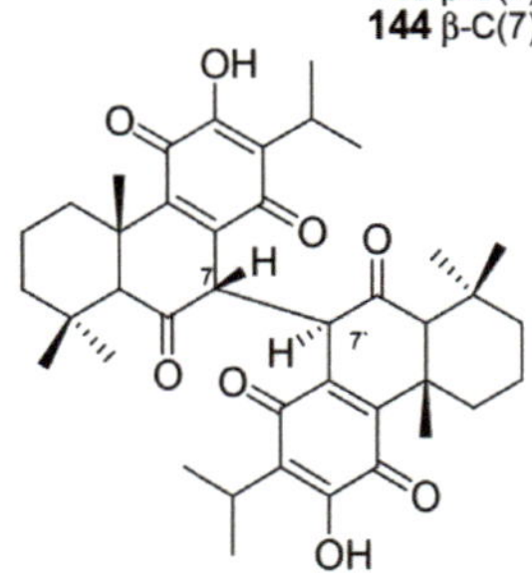

149

2.12.10. Plectranthus ornatus

Traditionally, the plants were used for treatment of stomach and liver diseases and as a substitute of *P. barbatus*. The phytochemical studies resulted in the isolation of 11 neoclerodanes (plectrornatins A (**160**) [75], 11*R**-acetoxykolavenic acid (**161**), 11*R**-acetoxy-2-oxokolavenic acid (**162**), 11*R**-acetoxy-3*β*-hydroxyneocleroda-4(18),13*E*-dien-15-oic acid (**163**) [76], ornatins A–E (**164-168**), 3*β*-hydroxyneocleroda-4(18),13*E*-dien-15-oic acid (**169**) [77]; 7 labdanes (plectrornatins B (**170**), C (**171**), [75],6-*O*-acetylforskolin (**172**); 1,6-di-*O*-acetylforskolin (**173**), 1,6-di-*O*-acetyl-9-deoxyforskolin (**174**) [76, 78], rhinocerotinoic acid (**175**) [66], 8*β*-hydroxylabd-13-en-15-oic acid (**176**) [77]); 2 abietanes (14-*O*-acetyl-coleon U (**177**), coleon R (**110**)) and a halimane derivative, (11R*-acetoxyhalima-5,13E-dien-15-oic acid (**178**) [79]) in addition to *β*-sitosterol and stigmasterol, 3*β*-acetyl-*α*-amyrin, and friedelin. Inversion at C-13 of 1,6-di-*O*-acetyl-9-deoxyforskolin (**174**) was carried out based on correlations between ^{13}C NMR experimental data and HF/6-31G* calculation [80]. **160, 161** showed moderate antimicrobial. **178** exhibited growth inhibitory activity against five *Staphylococcus* and five *Enterococcus* strains [75]. Ornatin C, D, E and three related diterpenes displayed marginal bactericidal or bacteriostatic effects against the Gram-positive strains [77].

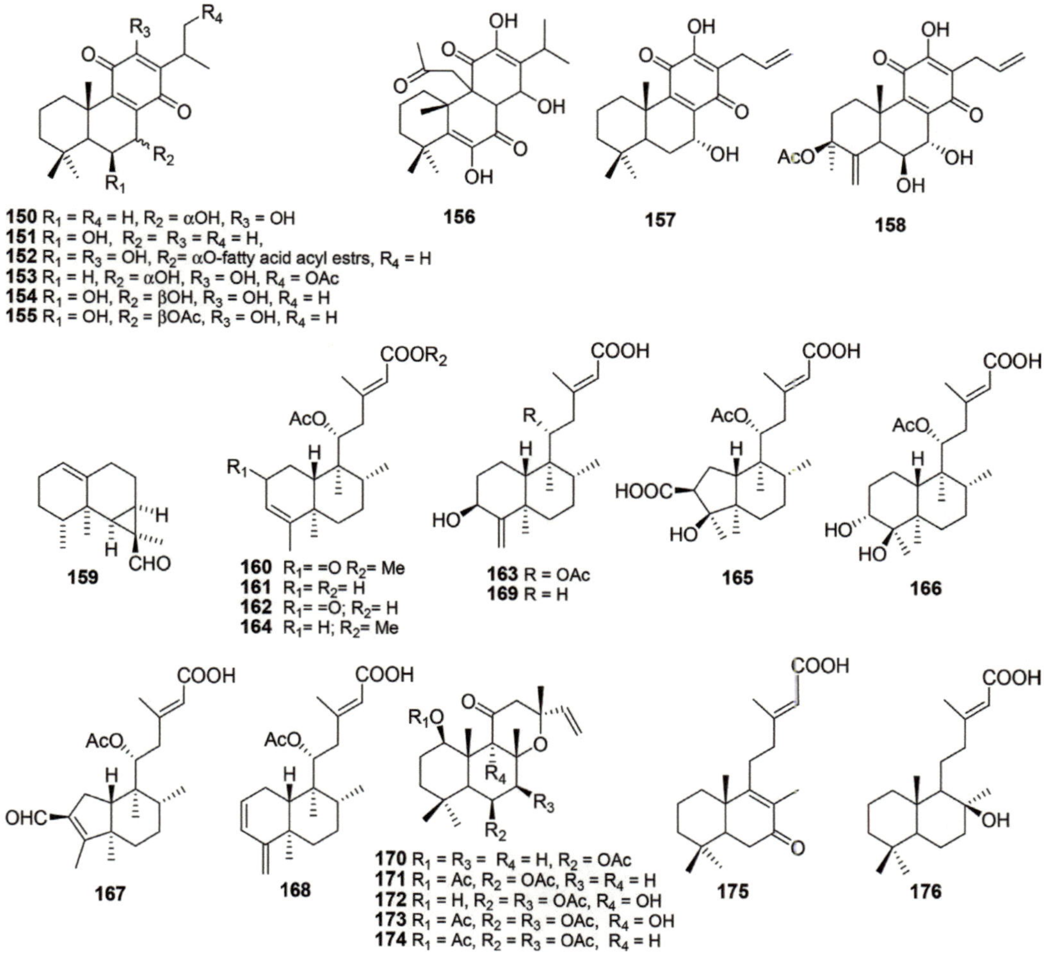

2.12.11. *Plectranthus porcatus*

(13*S*,15*S*)-6*β*,7*α*,12*α*,19-tetrahydroxy-13*β*,16-cyclo-8-abietene-11,14-dione (**179**) has been isolated and showed weak antibacterial activity against *S. aureus* [81].

2.12.12. *Plectranthus saccatus*

Ent-7*α*-acetoxy-15-beyeren-18-oic acid (**180**), *ent*-3*β*-(3-methyl-2-butenoyl) oxy-15-beyeren-19-oic acid (**181**), and ent-3*β*-(3-methylbutanoyl) oxy-15-beyeren-19-oic acid (**182**). Both **181** and **182** showed insect antifeedant activity against *Spodopteralittoralis*, while **180** showed no antibacterial activity [81, 82].

2.12.13. Plectranthus strigosus

9 abietanes (parviflorones A (**183**), B (**184**), C (**185**), D (**114**), E (**186**), F (**115**), G (**187**), and H (**188**) [83], and hinokiol (**189**)) [84]), 3 kauranes (*ent*-16-kauren-19-ol (**190**), *ent*-16-kauren-19-oic acid (**191**), xylopic acid (**192**), xylopinic acid (**193**)), and 2 sesquiterpens (4*β*,6*β*-dihydroxy-1*α*,5*β*(H)-guai-9-ene (**194**) 4*β*,6*β*-dihydroxy-1*α*,5*β*(H)-guai-10(14)-ene (**195**)), were isolated [84]. A bioactivity study revealed herpetic inhibitory properties for (**190**) and (**191**) [84].

2.13. *Salvia* genus

The genus *Salvia* is known as sage and is the largest genus in Lamiaceae, comprising over 900 species distributed throughout the world. *Salvia is* represented by 30 species in SA, distributed mainly in great cape region. The chemistry of *Salvia* is rich in diterpenoids and different skeletons have been reported, also, many members of this genus is well known for its curative and medicinal properties like *S. officinalis* and *S. miltiorrhiza.*

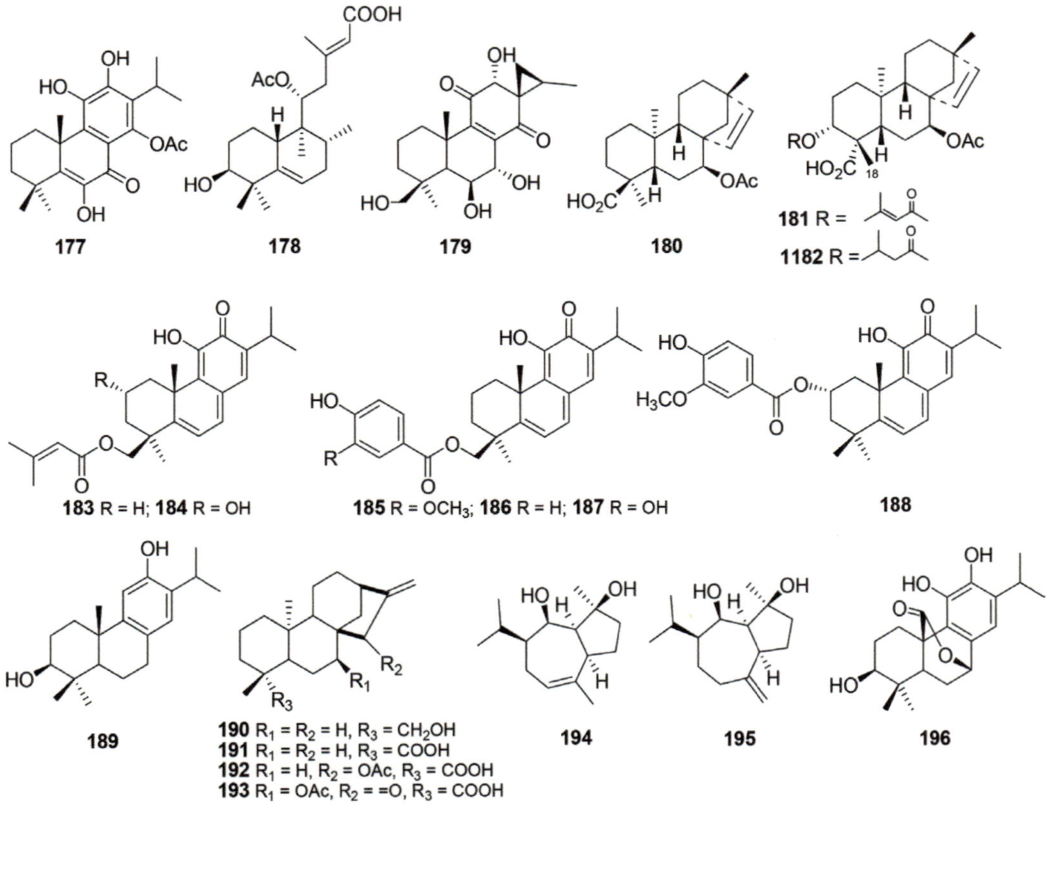

2.13.1. *Salvia africana-lutea*

Carnosol (**196**), rosmadial (**197**), and carnosic acid (**198**-characterized as its methyl ester) were isolated. Compound **198** exhibited potent activity against *M. tuberculosis* and cytotoxic activity against a breast (MCF-7) human cancer cell line [45].

2.13.2. *Salvia chamelaeagnea*

Four compounds were isolated: carnosol (**196**), 7-O-methylepirosmanol (**200**), oleanolic and ursolic acids as the active principles against *S. aureus* [85].

2.13.3. *Salvia coccinea*

Momordic acid, methyl ester (**201**) [86], salviacoccin (**202**) [87], dehydrouvaol (**203**), and uvaol (**204**) [88] were isolated.

2.13.4. *Salvia disermas*

The aerial parts afforded ocotillol II (**205**) [89].

2.13.5. *Salvia radula*

Betulafolientriol oxide (**206**) was isolated [90].

2.13.6. *Salvia reflexa*

Four neoclerodanes were isolated and identified as salviarin (**207**), 6β-hydroxysalviarin(**208**), 15,16-epoxy-8α-hydroxyneocleroda-2,13(16),14-triene-17,12*R*:18,19-diolide (**209**), and 5,6-secoclerodane, 7,8-didehydrorhyacophiline (**210**) [91].

2.13.7. *Salvia repens*

The whole plant extract yielded 12-methoxycarnosic acid (**199**) with antiprotozoal activity against *Leishmania donovani* amastigotes and cytotoxicity against the L6-cells [92].

2.13.8. *Salvia verbenaca*

The plant yielded β-sitosterol, ursolic acid, dehydroursolic acid, sitosteryl-3-β-D-glucoside [93], taxodione (**211**), horminone (**150**) and 7α-acetoxy-6β-hydroxyroyleanone (**102**) [94], verbenacine (**212**) and salvinine (**213**) [95].

2.14. *Solenostemon* genus

Solenostemon genus is from *S. rotundifolius*; oleanolic acid was isolated as a major component [96].

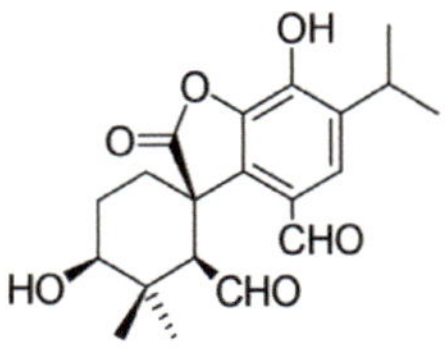

197

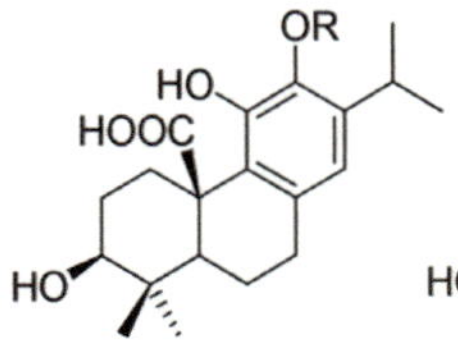

198 R= H; **199** R = Me

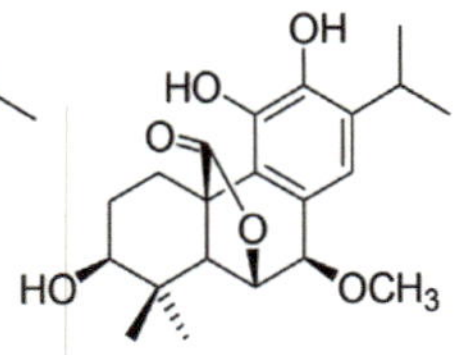

200

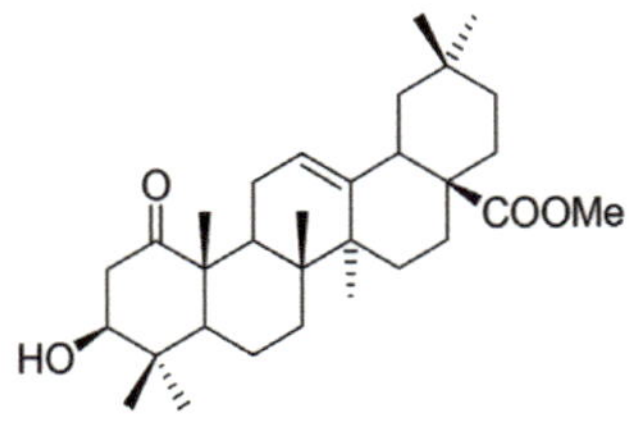

201

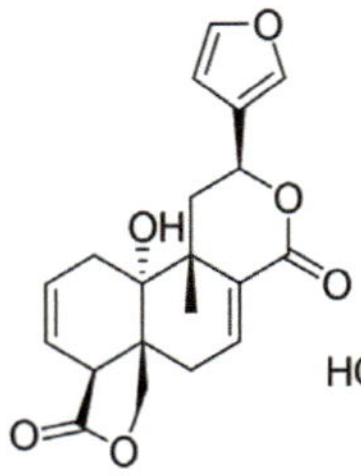

202

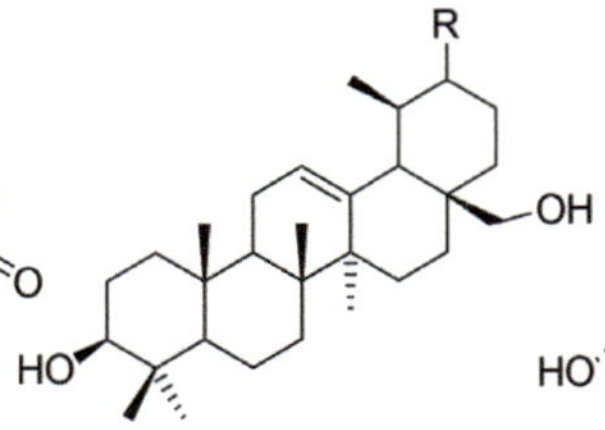

203 R = =CH$_2$
204 R = α-CH$_3$

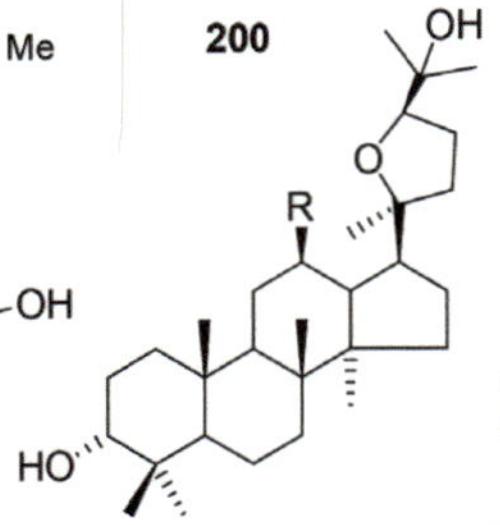

205 R = H
206 R = OH

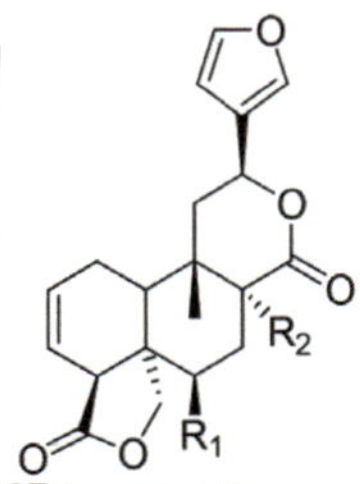

207 R$_1$ = R$_2$ = H
208 R$_1$ = OH, R$_2$ = H
209 R$_1$ = H, R$_2$ = OH

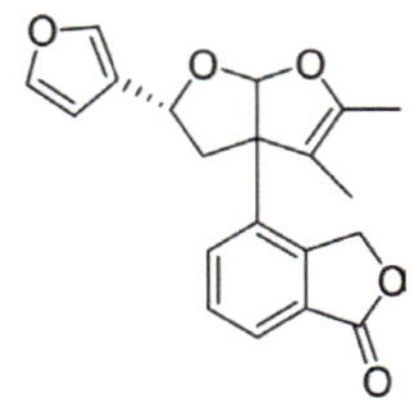

210

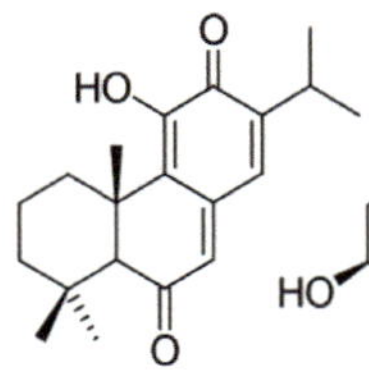

211

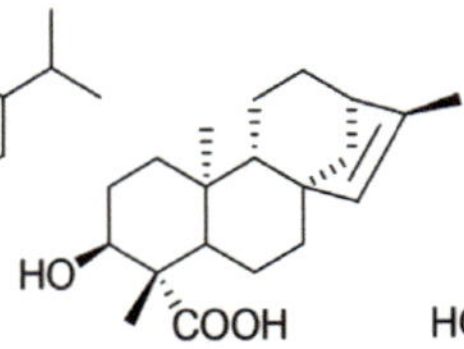

212

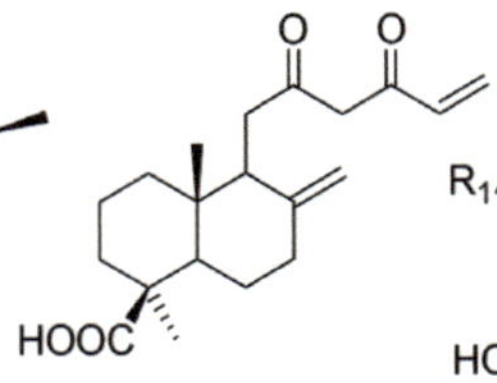

213

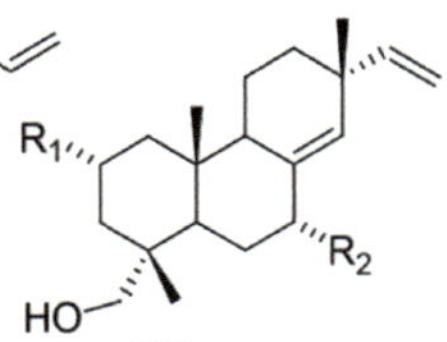

214 R$_1$ = H, R$_2$ = OH
215 R$_1$ = OH, R$_2$ = H

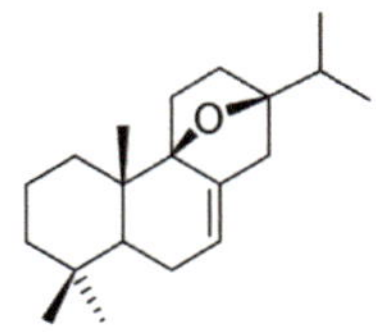

216

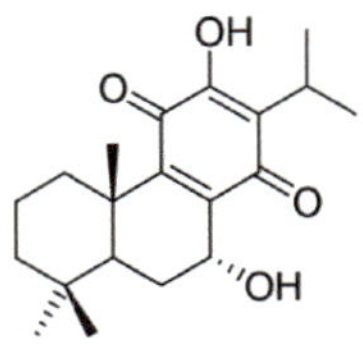

217

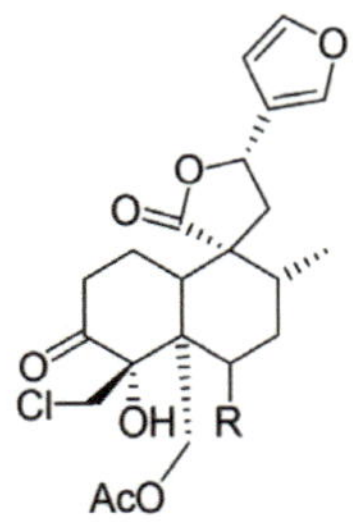

218 R = =O
219 R = OAc

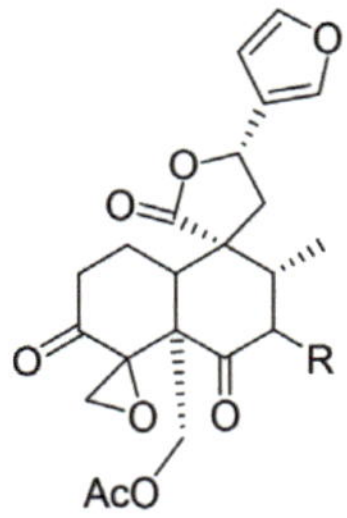

220 R = OH
221 R = H

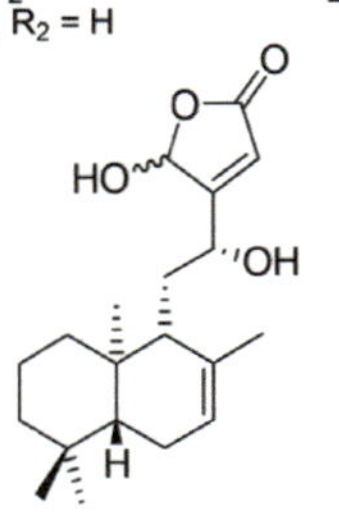

222

2.15. *Tetradenia* genus

Seven species were recorded in SA, one of them *T. riparia* is widely distributed in Africa and showed interesting chemical profile. Several compounds have been isolated from the leaves of this plant, including 8(14),15-sandaracopimaradiene-7α,18-diol (**214**) [97], 8(14), 15-sandaracopimaradiene-2α,18-diol (**215**) [98], 9β,13β-epoxy-7-abietene (**216**), 6,7-dehydroroyleanone (**136**) [99], and ibozol (**217**) [100].

Compound (**214**) exhibited antimicrobial activity (**213**). Compound (**215**) showed papaverine-like antispasmodic activity on guinea pig ileum contracted by methacholine, histamine, or $BaCl_2$ and on the noradrenaline-induced contractions of rabbit aorta [101]. It also showed activities against *Trichomonasvulgaris* with MIC of 20–40 μg/mL [102], wheat rootlets inhibition activity (MIC7.81 μg/mL) [103], and *M. tuberculosis*[104].

2.16. *Teucrium* genus

Three species were recorded in SA. From *T. africanum*tafricanins A (**218**) and B(**219**), teutrifidin (**220**) and 4α,18-epoxytafricanin A (**221**) were isolated [105].

2.17. *Vitex* genus

Vitex genus is represented by 12 species in SA. The fraction responsible for antimicrobial activity of *V. rehmannii* was purified to give a labdane diterpene as an inseparable epimeric mixture of 12*S*,16*S*/*R*-dihydroxy-*ent*-labda-7,13-dien-15,16-olide (**222**). The extract and the labdane diterpene exhibited good antimalarial activity, with the labdane diterpene being the most active IC_{50}: 2.39 ± 0.64 μg/mL [106].

3. Conclusion

South African flora characterized by high endemism and unique floral kingdom is only located in the great cape region. Lamiaceae is represented by ~308 species widely distributed all over the country. In general, the bioprospecting of SA flora including Lamiaceae is not reached; yet the required level and more attention are required to explore the potential of their chemical constituents. The present work shades the light on the isolated terpenoids of all listed species in updated SA flora checklist. It is interesting to indicate that *Plectranthus* genus contains mostly abietane diterpenes and shows potent activity as demonstrated by coleon U and parviflorons F and D. On the other hand, leoleorin C from *L. Leonurus* showed moderate binding affinity (*Ki* = 2.9 μM) to the Sigma 1 receptor. These compounds and others may be considered as a model for drug discovery for human benefits.

Acknowledgements

Cape Peninsula University of Technology and National Research Foundation, South Africa, (grant No. 106055).

Conflict of interest

The author declares no conflict of interest to disclose.

Author details

Ahmed A. Hussein

Address all correspondence to: mohammedam@cput.ac.za

Chemistry Department, Cape peninsula University of Technology, Bellville Campus, Western Cape, South Africa

References

[1] Domitilla R. The Red List of South African plants—A global first. African Journal of Science. 2011;**107**:3-4. DOI: 10.4102/sajs.v107i3/4.653

[2] Domitilla CR, von Staden L, John SD. Lessons from the conservation assessment of the South African mega flora. Annals of the Missouri Botanical Garden. 2013;**99**:221-230. DOI: 10.3417/2011111

[3] Heywoo VH. Flowering Plants of the World. Oxford: Oxford University Press; 1978. pp. 239-240

[4] Lee CJ, Chen LJ, Chang TL, Ke WM, Lo YF, Wang CC. The correlation between skin-care effects and phytochemical contents in *Lamiaceae* plants. Food Chemistry. 2011;**124**:833-841

[5] Khaled-Khodja N, Boulekbache-Makhlouf L, Madan K. Phytochemical screening of antioxidant and antibacterial activities of methanolic extracts of some *Lamiaceae*. Industrial Crops and Products. 2014;**61**:41-48

[6] Ramos M, Jimènez A, Peltzer M, Garrigòs MC. Characterization and antimicrobial activity studies of polypropylene films with carvacrol and thymol for active packaging. Journal of Food Engineering. 2012;**109**:513-519

[7] Klopper RR, Chatelain C, Bänninger V, Habashi C, Steyn, HM, De Wet BC, Arnold TH, Gautier L, Smith GF, Spichiger R. Checklist of the flowering plants of Sub-Saharan Africa. An index of accepted names and synonyms. Southern African Botanical Diversity Network Report No. 42. SABONET, Pretoria; 2006

[8] https://peerj.com/preprints/2277v1.pdf

[9] http://posa.sanbi.org/searchspp.php [Accessed: January 2018]

[10] https://www.cas.org/products/scifinder [Accessed: November–December 2017]

[11] Alasbahi RH, Melzig MF. *Plectranthus barbatus*: A review of phytochemistry, ethnobotanical uses and pharmacology—Part 2. Planta Medica. 2010;**76**:753-765. DOI: 10.1055/s-0029-1240898

[12] Alasbahi RH, Melzig MF. *Plectranthus barbatus*: A review of phytochemistry, ethnobotanical uses and pharmacology—Part 1. Planta Medica. 2010;**76**:653-661. DOI: 10.1055/s-0029-1240898

[13] Batista O, Duarte A, Nascimento J, Simoes MF, de laTorre MC, Rodriguez B. Structure and antimicrobial activity of diterpenes from the roots of *Plectranthus hereroensis*. Journal of Natural Products. 1994;**57**:588-561. DOI: 10.1021/np50108a031

[14] Dellar JE, Cole MD, Waterman PG. Unusual antimicrobial compounds from *Aeollanthus buchnerianus*. Experientia. 1996;**52**:175-179

[15] Rijo P, Simoes M-F, Duarte A, Rodriguez B. Isopimarane diterpenoids from *Aeollanthus rydingianus* and their antimicrobial activity. Phytochemistry. 2009;**70**:1161-1165. DOI: 10.1016/j.phytochem.2009.06.008

[16] Davies-Coleman MT, Rivett DE. Transformation of hispanolone from *Ballota africana* into 15,16-epoxy-9-hydroxylabda-13(16),14-diene. South African Journal of Chemistry. 1990;**43**:117-119

[17] Carreiras MC, Rodriguez B, Lopez-Garcia RE, Rabanal RM. A dimer of d-pinocarvone from *Cedronella canariensis*. Phytochemistry. 1987;**26**:3351-3353

[18] Masevhe NA, Awouafack MD, Ahmed AS, McGaw LJ, Eloff JN. Clerodendrumic acid, a new triterpenoid from *clerodendrum glabrum* (verbenaceae), and antimicrobial activities of fractions and constituents. Helvetica Chimica Acta. 2013;**96**:1693-1703. DOI: 10.1002/hlca.201200552

[19] Mujovo SF, Hussein AA, Meyer JJM, Fourie B, Muthivhi T, Lall N. Bioactive compounds from *Lippia javanica* and *Hoslundia opposite*. Natural Product Research. 2008;**22**:1047-1054. DOI: 10.1080/14786410802250037

[20] Annan K, Jackson N, Dickson RA, Sam GH, Komlaga G. Acaricidal effect of an isolate from *Hoslundia opposite* vahl against *Amblyomma variegatum* (Acari: Ixodidae). Pharmacognosy Research. 2011;**3**:185-188

[21] Achenbach H, Waibel R, Nkunya MHH, Weenen H. Antimalarial compounds from *Hoslundia opposite*. Phytochemistry. 1992;**31**:3781-3784

[22] Perptiseda-Miranda R, Gascon-Figueroa M. Chemistry of *Hyptis mutabilis*: New penta cyclic triterpenoids. Journal of Natural Products. 1988;**51**(5):996-998. DOI: 10.1021/np50059a035

[23] Fragoso-Serrano M, Gonzalez-Chimeo E, Pereda-Miranda R. Novel labdane diterpenes from the insecticidal plant *Hyptis spicigera*. Journal of Natural Products. 1999;**62**:45-50. DOI: 10.1021/NP980222Z

[24] Piozzi F, Bruno M, Rosselli S, Maggio A. Structure and biological activity of the furan-diterpenoids from the genera *Leonotis* and *Leonurus*. Heterocycles. 2007;**74**:31-52

[25] McKenzie JM, Green IR, Mugabo P. Leonurun, a novel labdane diterpenoid from *Leonotis leonurus*. South African Journal of Chemistry. 2006;**59**:114-116

[26] Obikeze KC, McKenzie JM, Green IR, Mugabo P. Characterization and cardiovascular effects of (13S)-9α,13α-epoxylabda-6β(19),15(14)dioldilactone, a diterpenoid isolated from *Leonotis leonurus*. South African Journal of Chemistry. 2008;**61**:119-122

[27] Naidoo D, Maharaj V, Crouch NR, Ngwane A. New labdane-type diterpenoids from *Leonotis leonurus* support circumscription of *Lamiaceae*. Biochemical Systematics and Ecology. 2011;**39**:216-219. DOI: 10.1016/j.bse.2010.12.021

[28] Fang H, Charlotte L, Wayne WH. Leonurenones A–C: Labdane diterpenes from *Leonotis leonurus*. Phytochemistry. 2012;**83**:168-172. DOI: 10.1016/j.phytochem.2012.07.014

[29] Wu H, Li J, Fronczek FR, Ferreira D, Burandt CL Jr, Setola V, Roth BL, Zjawiony JK. Labdane diterpenoids from *Leonotis leonurus*. Phytochemistry. 2013;**91**:229-235. DOI: 10.1016/j.phytochem.2012.02.021

[30] Agnihotri VK, El Sohly HN, Smillie TJ, Khan IA, Walker LA. Constituents of *Leonotis leonurus* flowering tops. Phytochemistry Letters. 2009;**2**:103-105. DOI: 10.1016/j.phytol.2009.02.001

[31] White JD, Manchand PS, Whalley WB. The structure of leonotin, a novel furanoid diterpene. Journal of Chemical Society D. 1969:0,1315-1316. DOI: 10.1039/C29690001315

[32] White JD, Manchand PS, Whalley WB. Structure of nepetaefolin, a prefuranoid diterpene. Journal of the American Chemical Society. 1972;**92**:5527-5528. DOI: 10.1021/ja00721a046

[33] Blount JF, Manchand PS. X-ray structure determination of methoxynepetaefolin and nepetaefolinol, labdane diterpenoids from *Leonotis nepetaefolia* R.Br. Journal of the Chemical Society, Perkin Transactions 1: Organic and Bio-Organic Chemistry. 1980; **1**:264-268. DOI: 10.1039/P19800000264

[34] Manchand PS. Methoxyneptefolin, a new labdane diterpene from *Leonotis nepetaefolia*. Tetrahedron Letters. 1973;**21**:1907-1908. DOI: 10.1016/S0040-4039(01)96273-2

[35] Purushothaman KK, Vasanth S, Connolly JD. Nepetefolinol and two related diterpenoids from *Leonotis nepetaefolia*. Journal of the Chemical Society, Perkin Transactions 1: Organic and Bio-Organic Chemistry. 1974;**23**:2661-2663. DOI: 10.1039/p19740002661

[36] Von Dreele RB, Pettit GR, Ode RH, Perdue RE, Jr, White JD, Manchand PS. The crystal and molecular structure of the unusual spiro dihydrofuran diterpene nepetaefolin. Journal of the American Chemical Society. 1975;**97**:6236-6240. DOI: 10.1021/ja00854a049

[37] Govindasamy L, Rajakannan V, Velmurugan D, Banumathi S, Vasanth S. Structural studies on three plant diterpenoids from *Leonotis nepetaefolia*. Crystal Research and Technology. 2002;**37**:896-909. DOI: 10.1002/1521-4079(200208)37:8<896:AID-CRAT896>3.0.CO;2-F

[38] Boalino DM, Tinto WF. A new diterpene of *Leonotis nepetaefolia*. Heterocycles. 2004; **63**:383-387

[39] Li J, Fronczek FR, Ferreira D, Burandt CL, Setola V, Roth BL, Zjawiony JK. Bis-spirolabdane diterpenoids from *Leonotis nepetaefolia*. Journal of Natural Products. 2012;**75**:728-734. DOI: 10.1021/np3000156

[40] Takeda T, Narukawa Y, Hada N. Studies on the constituents of *Leonotis nepetaefolia*. Chemical and Pharmaceutical Bulletin. 1999;**47**:284-286. DOI: 10.1248/cpb.47.284

[41] Habtemariam S, Gray AI, Waterman PG. Diterpenes from the leaves of *Leonotis ocymifolia* var. *Raineriana*. Journal of Natural Products. 1994;**57**:1570-1574. DOI: 10.1021/np50113a017

[42] Hussein AA, Meyer MJJ, Rodriguez B. Complete ^{1}H and ^{13}C NMR assignments of three labdane diterpenoids isolated from *Leonotis ocymifolia* and six other related compounds. Magnetic Resonance in Chemistry. 2003;**41**:147-151

[43] Rahman MM, Gibbons S. Antibacterial constituents of *Neohyptis paniculata*. Fitoterapia. 2015;**105**:269-272. DOI: 10.1016/j.fitote.2015.07.012

[44] Upadhyay RK, Misra LN, Singh G. Sesquiterpene alcohols of the copane series from essential oil of *Ocimum americanum*. Phytochemistry. 1991;**30**(2):691-693. DOI: 10.1016/0031-9422(91)83755-A

[45] Hussein AA, Meyer JJM, Jimeno ML, Rodriguez B. Bioactive diterpenes from *Orthosiphon labiatus* and *Salvia africana-lutea*. Journal of Natural Products. 2007;**70**(2):293-295. DOI: 10.1021/np0680376

[46] Kapewangolo P, Omolo JJ, Bruwer R, Fonteh P, Meyer D. Antioxidant and anti-inflammatory activity of *Ocimum labiatum* extract and isolated labdane diterpenoid. Journal of Inflammation (London United Kingdom). 2015;**12**:1-28

[47] Rasamiravaka T, Jaziri M, Rasamiravaka T, Ngezahayo J, Pottier L, Oliveira RS, Florence S, Caroline S, Pierre D, Jeremie N, et al. Terpenoids from *Platostoma rotundifolium* (Briq.) A. J. Paton Alter the expression of quorum sensing-related virulence factors and the formation of biofilm in *Pseudomonas aeruginosa* PAO1. International Journal of Molecular Sciences. 2017;**18**(6):1270

[48] Liu G, Ruedi P. Phyllocladanes (13β-kauranes) from *Plectranthus ambiguus*. Phytochemistry. 1996;**41**:1563-1568. DOI: 10.1016/0031-9422(95)00816-0

[49] Liu G, Muller R, Ruedi P. Chemical transformations of phyllocladane (=13β-kaurane) diterpenoids. Helvetica Chimica Acta. 2003;**86**:420-438. DOI: 10.1002/hlca.200390043

[50] Chen Y-S, Yu H-M, Shie J-J, Cheng T-JR, Wu C-Y, Fang J-M, Wong C-H. Chemical constituents of *Plectranthus amboinicus* and the synthetic analogs possessing anti-inflammatory activity. Bioorganic and Medicinal Chemistry. 2014;**22**:1766-1772. DOI: 10.1016/j.bmc.2014.01.009

[51] Yulianto W, Andarwulan N, Giriwono PE, Pamungkas J. HPLC-based metabolomics to identify cytotoxic compounds from *Plectranthus amboinicus* (Lour.) Spreng against human breast cancer MCF-7Cells. Journal of Chromatography B: Analytical Technologies in the Biomedical and Life Sciences. 2016;**1039**:28-34. DOI: 10.1016/j.jchromb.2016.10.024

[52] Arihara S, Ruedi P, Eugster CH. New spirocyclopropylcyclohexenedione diterpene type compounds.Coleons M, N, P, Q, and R and barbatusin from *Plectranthus caninus* and coleon O from *Coleus somaliensis*. Helvetica Chimica Acta. 1975;**58**:343-356. DOI: 10.1002/hlca.19750580203

[53] Arihara S, Ruedi P, Eugster CH. Diterpenoid leaf-gland pigments:coleons S and T from *Plectranthus caninus* Roth (Labiatae), a new diosphenol-*trans*-A/B-6,7-diketone pair of the abietane series. Helvetica Chimica Acta. 1977;**60**:1443-1447. DOI: 10.1002/hlca.19770600436

[54] Uchida M, Ruedi P, Eugster CH. Leaf-gland pigments from Labiatae:ecklonoquinones A and B, two novel dibenzo-*p*-dioxin-*O*-quinones from *Plectranthus ecklonii* Benth. Helvetica Chimica Acta. 1980;**63**:225-231. DOI: 10.1002/hlca.19800630122

[55] Nyila MA, Leonard CM, Hussein AA, Lall N. Bioactivities of *Plectranthus ecklonii* constituents. Natural Product Communications. 2009;**4**:1177-1180

[56] Burmistrova O, Perdomo J, Simoes MF, Rijo P, Quintana J, Estevez F. The abietane diterpenoid parvifloron D from *Plectranthus ecklonii* is a potent apoptotic inducer in human leukemia cells. Phytomedicine. 2015;**22**:1009-1016. DOI: 10.1016/j.phymed.2015.06.013

[57] Stavri M, Paton A, Skelton BW, Gibbons S. Antibacterial diterpenes from *Plectranthus ernstii*. Journal of Natural Products. 2009;**72**:1191-1194. DOI: 10.1021/np800581s

[58] Gaspar-Marques C, Simoes MF, Duarte A, Rodriguez B. Labdane and kaurane diterpenoids from *Plectranthus fruticosus*. Journal of Natural Products. 2003;**66**(4):491-496. DOI: 10.1021/np020493g

[59] Gaspar-Marques C, Simoes MF, Rodriguez B. Further labdane and kaurane diterpenoids and other constituents from *Plectranthus fruticosus*. Journal of Natural Products. 2004;**67**:614-621. DOI: 10.1021/np030490j

[60] Purushothaman KK, Sarada A, Saraswathy A. Some rare chemical constituents of a traditional drug of doubtful origin. Indian Drugs. 1986;**23**:579-580

[61] Uchida M, Miyase T, Yoshizaki F, Bieri JH, Ruedi P, Eugster CH. 14-Hydroxytaxodione as major diterpenoid in *Plectranthus grandidentatus* Gurke; isolation of seven new dimeric diterpenoids from *P. grandidentatus*, *P. myrianthus* Briq. and *Coleus carnosus* Hassk.:Structures of grandidones A, 7-epi-A, B, 7-epi-B, C, D and 7-epi-D. Helvetica Chimica Acta. 1981;**64**:2227-2250. DOI: 10.1002/hlca.19810640729

[62] Antonio P, Batista O, Simoes MF, Nascimento J, Duarte A, de la Torre MC, Rodriguez B. Abietane diterpenoids from *Plectranthus grandidentatus*. Phytochemistry. 1996;**44**: 325-327. DOI: 10.1016/S0031-9422(96)00467-0

[63] Marque CC, Pedro M, Simoes MFA, Nascimento MSJ, Pinto MMM, Rodriguez B. Effect of abietane diterpenes from *Plectranthus grandidentatus* on the growth of human cancer cell lines. Planta Medica. 2002;**68**:839-840. DOI: 10.1055/s-2002-34407

[64] Cerqueira F, Cordeiro-Da-Silva A, Gaspar-Marques C, Simoes F, Pinto MMM, Nascimento MSJ. Effect of abietane diterpenes from *Plectranthus grandidentatus* on T-and B-lymphocyte proliferation. Bioorganic and Medicinal Chemistry. 2004;**12**:217-223. DOI: 10.1016/j.bmc.2003.10.006

[65] Gaspar-Marques C, Simoes MF, Rodriguez B. A Trihomo abietane diterpenoid from *Plectranthus grandidentatus* and an unusual addition of acetone to the ortho-quinone system of crypto tanshinone. Journal of Natural Products. 2005;**68**:1408-1411. DOI: 10.1021/np0580457

[66] Gaspar-Marques C, Rijo P, Simoes MF, Duarte MA, Rodriguez B. Abietanes from *Plectranthus grandidentatus* and *P. hereroensis* against methicillin-and vancomycin-resistant bacteria. Phytomedicine. 2006;**13**:267-271. DOI: 10.1016/j.phymed.2005.06.002

[67] Rijo P, Gaspar-Marques C, Simoes MF, Jimeno ML, Rodriguez B. Further diterpenoids from *Plectranthus ornatus* and *P. grandidentatus*. Biochemical Systematics and Ecology. 2007;**35**:215-221. DOI: 10.1016/j.bse.2006.10.011

[68] Coutinho I, Pereira G, Simoes MF, Corte-Real M, Goncalves J, Saraiva L. Selective activation of protein kinase C-δ and -ε by 6, 11,12,14-tetrahydroxy-abieta-5,8,11,13-tetraene-7-one (coleon U). Biochemical Pharmacology. 2009;**78**:449-459. DOI: 10.1016/j.bcp.2009.04.026

[69] Burmistrova O, Simoes MF, Rijo P, Quintana J, Bermejo J, Estevez F. Antiproliferative activity of abietane diterpenoids against human tumor cells. Journal of Natural Products. 2013;**76**:1413-1423. DOI: 10.1021/np400172k

[70] Rijo P, Duarte A, Francisco AP, Semedo-Lemsaddek T, Simoes MF. *In vitro* antimicrobial activity of royleanone derivatives against gram-positive bacterial pathogens. Phytotherapy Research. 2014;**28**:76-81. DOI: 10.1002/ptr.4961

[71] Batista O, Simoes MF, Nascimento J, Riberio S, Duarte A, Rodriguez B, de la Torre MC. A rearranged abietane diterpenoid from *Plectranthus hereroensis*. Phytochemistry. 1996; **41**:571-573. DOI: 10.1016/0031-9422(95)00646-X

[72] Rodriguez B, de laTorre MC, Simoes F, Batista O, Nascimento J, Duarte A, Mayer R. Revision of the structure of an aristolane sesquiterpene aldehyde isolated from the root of *Plectranthus hereroensis* and *Aristolochia debilis*. Phytochemistry. 1995;**38**:905-907. DOI: 10.1016/0031-9422(94)00741-B

[73] Batista O, Simoes MF, Duarte A, Valdeira ML, de laTorre MC, Rodriguez B. An antimicrobial abietane from the root of *Plectranthus hereroensis*. Phytochemistry. 1995;**38**: 167-169. DOI: 10.1016/0031-9422(94)00586-I

[74] Kubinova R, Porizkova R, Navratilova A, Farsa O, Hanakova Z, Bacinska A, Cizek A, Valentova M. Antimicrobial and enzyme inhibitory activities of the constituents of *Plectranthus madagascariensis* (Pers.) Benth. Journal of Enzyme Inhibition and Medicinal Chemistry. 2014;**29**:749-752. DOI: 10.3109/14756366.2013.848204

[75] Rijo P, Gaspar-Marques C, Simoes MF, Duarte A, Apreda-Rojas MC, Cano FH, Rodriguez B. Neoclerodane and labdane diterpenoids from *Plectranthus ornatus*. Journal of Natural Products. 2002;**65**:1387-1390. DOI: 10.1021/np020203w

[76] Oliveira PM, Ferreira AA, Silveira D, Alves RB, Rodrigues GV, Emerenciano VP, Raslan DS. Diterpenoids from the aerial parts of *Plectranthus ornatus*. Journal of Natural Products. 2005;**68**:588-591. DOI: 10.1021/np049827

[77] Avila FN, Pinto FCL, Sousa TS, Torres MCM, Costa-Lotufo LV, Rocha DD, de Vasconcelos MA, Cardoso-Sa N, Teixeira EH, Albuquerque MRJR, et al. Miscellaneous diterpenes from the aerial parts of *Plectranthus ornatus* codd. Journal of the Brazilian Chemical Society. 2017;**28**:1014-1022. DOI: 10.21577/0103-5053.20160255

[78] Rijo P, Simoes MF, Rodriguez B. Structural and spectral assignment of three forskolin-like diterpenoids isolated from *Plectranthus ornatus*. Magnetic Resonance in Chemistry. 2005;**43**:595-598. DOI: 10.1002/mrc.1600/pdf

[79] Rijo P, Rodriguez B, Duarte A, Simoes MF. Antimicrobial properties of *Plectranthus ornatus* extracts, 11-acetoxyhalima-5,13-dien-15-oic acid metabolite and its derivatives. Natural Products Journal. 2011;**1**:57-64. DOI: 10.2174/2210315511101010057

[80] Oliveira PM, Pacheco AG, Alves RB, Pilo-Veloso D, Raslan DS, de Carvalho Alcantara AF. A new configurational analysis of 1,6,7-triacetoxy-8,13-epoxy-14-labden-11-one isolated from *Plectranthus ornatus* based on NMR and theoretical calculations. Open Natural Products Journal. 2009;**2**:1-5

[81] Simoes MF, Rijo P, Duarte A, Barbosa D, Matias D, Delgado J, Cirilo N, Rodriguez B. Two new diterpenoids from *Plectranthus* species. Phytochemistry Letters. 2010;**3**:221-225. DOI: 10.1016/j.phytol.2010.08.002

[82] Wellsow J, Grayer RJ, Veitch NC, Kokubun T, Lelli R, Kite GC, Simmonds MSJ. Insect-antifeedant and antibacterial activity of diterpenoids from species of *Plectranthus*. Phytochemistry. 2006;**67**:1818-1825. DOI: 10.1016/j.phytochem.2006.02.018

[83] Alder AC, Ruedi P, Eugster CH. Glandular pigments from tropical labiates: Parviflorones from *Plectranthus strigosus* Benth. Helvetica Chimica Acta. 1984;**67**:1531-1534. DOI: 10.1002/hlca.19840670617

[84] Gaspar-Marques C, Simoes MF, Valdeira ML, Rodriguez B. Terpenoids and phenolics from *Plectranthus strigosus*, bioactivity screening. Natural Product Research. 2008;**22**:167-177. DOI: 10.1080/14786410701654560

[85] Kamatou GPP, Van Vuuren SF, Van Heerden FR, Seaman T, Viljoen AM. Antibacterial and antimycobacterial activities of South African *Salvia* species and isolated compounds from *S. chamelaeagnea*. South African Journal of Botany. 2007;**73**:552-557

[86] Mukherjee KS, Mukhopadhyay B, Brahmachar G. Anew triterpene from *Salvia coccinea*. Journal of the Indian Chemical Society. 2004;**81**:82-83

[87] Savona G, Bruno M, Paternostro M, Marco JL, Rodriguez B. Salviacoccin, a neoclerodane diterpenoid from *Salvia coccinea*. Phytochemistry. 1982;**21**:2563-2566

[88] Mukherjee KS, Ghosh PK. Further studies on *Salvia coccinea* Linn. Journal of the Indian Chemical Society. 1978;**55**:850

[89] Hawas UW, Gamal-Eldee AM, El-Toumy SAA, Meyer JJM, Hussein AA. Inhibition of the initiation stage of carcinogenesis by *Salvia disermas* constituents. Zeitschrift fuer Naturforschung, C: Journal of Biosciences. 2009;**64**:831-839. DOI: 10.1515/znc-2009-11-1213

[90] Kamatou GPP, van Zyl L, David H, van Heerden FR, Lourens ACU, Viljoen AM. Anti-malarial and anticancer activities of selected South African *Salvia* species and isolated compounds from *S. radula*. South African Journal of Botany. 2008;**74**:238-243. DOI: 10.1016/j.sajb.2007.08.001

[91] Nieto M, Gallardo VO, Rossomando PC, Tonn CE. 8-Hydroxysalviarin and 7,8-didehydrorhyacophiline, two new diterpenes from *Salvia reflexa*. Journal of Natural Products. 1996;**59**:880-882

[92] Mokoka TA, Peter XK, Fouche G, Moodley N, Adams M, Hamburger M, Kaiser M, Brun R, Maharaj V, Koorbanally N. Antileishmanial activity of 12-methoxycarnosic acid from *Salvia repens* Burch. ex Benth (Lamiaceae). South African Journal of Botany. 2014;**90**:93-95

[93] Saleh MRI, Nazmi SN. Phytochemical study of *Salvia verbenaca* L. herb. Egyptian Journal of Pharmaceutical Sciences. 1982;**20**:411-415

[94] Sabri NN, Abou-Donia AA, Assad AM, Ghaz NM, El-Lakany AM, Tempesta MS, Sanson DR. Abietane diterpene quinones from roots of *Salvia verbenaca* and *S. lanigera*. Planta Medica. 1989;**55**:582

[95] Ahmed B, Al-Howiriny TA, Al-Rehaily AJ, Mossa JS. Verbenacine and salvinine: two new diterpenes from *Salvia verbenaca*. Zeitschrift fuer Naturforschung, C: Journal of Biosciences. 2004;**59**:9-14

[96] Michael TD, English RB, Rivett DEA. Synrotolide, an α-pyrone from *Syncolostemon rotundifolius*. Phytochemistry. 1987;**26**:1497-1499

[97] Van Puyvelde L, Nyirankuliza S, Panebianco R, Boily Y, Geizer I, Sebikali B, De Kimpe N, Schamp N. Active principles of *Tetradenia riparia*. I. Antimicrobial activity of 8(14),15-sandaracopimaradiene-7α,18-diol. Journal of Ethnopharmacology. 1986; **17**:269-275

[98] Van Puyvelde L, De Kimpe N, Borremans F, Zhang W, Schamp N. 8(14),15-Sandaracopimaradiene-2α,18-diol, a minor constituent of the Rwandese medicinal plant *Tetradenia riparia*. Phytochemistry. 1987;**26**:493-495

[99] Gazim ZC, Rodrigues F, Amorin ACL, de Rezende CM, Sokovic M, Tesevic V, Vuckovic I, Krstic G, Cortez LER, Colauto NB, et al. New natural diterpene-type abietane from *Tetradenia riparia* essential oil with cytotoxic and antioxidant activities. Molecules. 2014;**19**:514-524. DOI: 10.3390/molecules19010514

[100] Fernandez ACAM, Rosa MF, Fernandez CMM, Bortolucci WC, Melo UZ, Siqueira VLD, Cortez DAG, Goncalves JE, Linde GA, Gazim ZC. Antimicrobial and antioxidant activities of the extract and fractions of *Tetradenia riparia* (hochst.) codd (lamiaceae) leaves. Brazil. Current Microbiology. 2017;**74**:1453-1460. DOI: 10.1007/s00284-017-1340-9

[101] Van Puyvelde L, Lefebvre R, Mugabo P, De Kimpe N, Schamp N. Active principles of *Tetradenia riparia*; II. Antispasmodic activity of 8(14),15-sandaracopimaradiene-7α,18-diol. Planta Medica. 1987;**53**:156-158

[102] Hakizamungu E, Van Puyvelde L, Wery M, De Kimpe N, Schamp N. Active principles of *Tetradenia riparia* III. Anti-trichomonas activity of 8(14),15-sandaracopimaradiene-7α,18-diol. Phytotherapy Research. 1988;**2**:207-208

[103] Van Puyvelde L, De Kimpe N, Ayobangira FX, Costa J, Nshimyumukiza P, Boily Y, Hakizamungu E, Schamp N. Wheat rootlet growth inhibition test of Rwandese medicinal plants: Active principles of *Tetradenia riparia* and *Diplolophium africanum*. Journal of Ethnopharmacology. 1988;**24**:233-246

[104] Van Puyvelde L, Ntawukiliyayo JD, Portaels F, Hakizamungu E. In vitro inhibition of mycobacteria by Rwandese medicinal plants. Phytotherapy Research. 1994;**8**:65-69

[105] Hanson JR, Rivett DEA, Ley SV, Williams DJ. The x-ray structure and absolute configuration of insect antifeedant clerodane diterpenoids from *Teucrium africanum*. Journal of the Chemical Society, Perkin Transactions 1: Organic and Bio-Organic Chemistry. 1982;**4**:1005-1008

[106] Nyiligira E, Viljoen AM, Van Heerden FR, Van Zyl RL, Van Vuuren SF, Steenkamp PA. Phytochemistry and *in vitro* pharmacological activities of South African *Vitex* (Verbenaceae) species. Journal of Ethnopharmacology. 2008;**119**:680-685

Chapter 3

Terpenes as Potential Antimalarial Drugs

Heloisa Berti Gabriel, Rodrigo AC Sussmann,
Emila A Kimura,
Adriana Alejandra Marin Rodriguez,
Ignasi Bofill Verdaguer,
Gabriela Carolina Fernandes Leite and
Alejandro Miguel Katzin

Additional information is available at the end of the chapter

http://dx.doi.org/10.5772/intechopen.75108

Abstract

A fact which favors the increase in morbidity and mortality of malaria cases in the world is the resistance to chemotherapeutic agents that the parasite presents. Therefore, it is necessary to identify new potential targets specific to the parasite in order to be able to perform a rational planning. One target for the evaluation of potential antimalarial compounds is isoprenoid synthesis, which occurs via the 2-C-methyl-*d*-erythritol-4-phosphate pathway in *Plasmodium falciparum*. Several intermediaries and final products of this pathway were identified in the parasite and lead us to the conclusion that it is different from the vertebrate host. In this chapter, we describe the effect of some monoterpenes and sesquiterpenes on *Plasmodium falciparum* and *Plasmodium berghei* as potential antimalarial drugs.

Keywords: terpenes, malaria, *Plasmodium falciparum*, *Plasmodium berghei*, isoprenoid

1. Introduction

Malaria is one of the major threats to human health, affecting an estimated number of 216 million peoples in 2016 all over the world, leading to 445,000 deaths, mainly in the African continent [1]. The human malaria is caused by six different species of the genus *Plasmodium*, which are *P. falciparum*, *P. vivax*, *P. ovale*, *P. malariae*, *Plasmodium knowlesi* and *P. simium*, where the last one is exclusive to the Brazilian Atlantic Forest [2].

Plasmodium was first described in 1880 by Laveran, who observed it in human erythrocytes. In human *Plasmodium* genus, the life cycle is very similar between species, being characterized by a sexual phase in vector *Anopheles* and an asexual phase in the human host that can be divided into liver phase and intraerythrocytic phase [3] (**Figure 1**).

The cycle begins with the injection of infective sporozoites of salivary glands of *Anopheles* into the human host bloodstream. Once there, the sporozoites make their way into the liver, where they infect hepatocytes to start the first massive replication, an asexual process, known as exo-erythrocytic schizogony. After 9–16 days, those new-formed merozoites are then released into the bloodstream in hepatocyte-derived vesicles [4], called merosomes, to avoid its capture by Kupffer cell. This liver endures this phase differently across different species which can last an average of 6 days (*P. falciparum*), 10 days (*P. vivax*), or 15 days (*P. ovale* and *P. malariae*) [5].

In the bloodstream, the merosomes are then disrupted liberating the merozoites, each merozoite infecting a single red blood cell (RBC). The process of invasion is complex, relying on diverse cell machineries, which permit the parasite to attach, reorientate, and invade, forming the parasitophorous vacuole [6]. Once in the erythrocyte, the parasite starts its asexual division, passing through different stages. The early trophozoite, called "ring stage", starts to develop, enlarging to a mature trophozoite that has a high metabolic index. In the late stage, multiple nuclear divisions

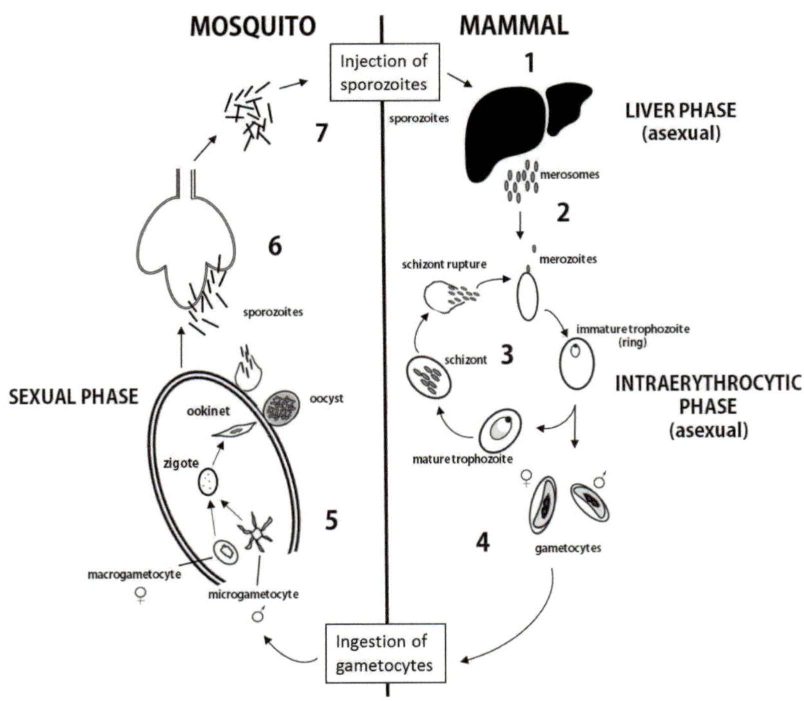

Figure 1. Malaria parasite life cycle. Schematic life cycle of *P. falciparum* in the invertebrate (left) and vertebrate hosts (right). 1. Hepatocytes invasion and exoerythrocytic schizogony to merozoites formation. 2. Release of merosomes in the blood stream. 3. Intraerythrocytic phase. 4. Parasite differentiation to gametocytes which ones could be ingested by invertebrate host. 5. Sexual phase in the midgut of invertebrate host. 6. Migration of sporozoites into salivary glands. 7. Injection of sporozoites during the blood meal.

are triggered without cytokinesis, forming schizonts. Each schizont holds an average of 32 merozoites (10 merozoites in average for *P. knowlesi* [7]) that are unleashed upon the RBC lysis. The whole process can take about 36–48 h in *P. falciparum*, 48 h in *P. vivax* and can reach even 72 h in *P. malariae*, but in *P. knowlesi* the cycle is 24 h, which is one factor that leads to its high virulence in humans [8]. Cell lysis coincides with fever symptoms, a response of immune system to the liberation of hemozoin and other parasite products into the bloodstream [9].

Within the red blood cells, the parasite can follow another path of development, differentiating into gametocytes. During a blood meal in an infected individual, the *Anopheles* female ingests those gametocytes. In the female of the *Anopheles* mosquito the parasite undergoes a meiotic division. Inside the mosquito gut, the gametocytes mature to form male and female gametes. The gametes undergo fertilization, forming the zygote, which transforms into an ookinete. The ookinete then penetrates the midgut and installs itself developing into oocyst. Under multiple cellular divisions, thousands of sporozoites are formed, which migrate to the salivary glands of the mosquito, to get expelled with anticoagulant factor contained in saliva during the next blood meal, restarting the cycle [10] (**Figure 1**).

Although *P. vivax* is the most prevalent parasite in the world, *P. falciparum* is responsible for most cases of severe malaria, being the most prevalent malarial parasite in the African continent, which accounts for 80% of the global disease burden. The groups with higher risk of malaria disease includes pregnant women, patients with HIV/AIDS, infants and children under 5 years old, whereas *P. falciparum* is responsible for about 70% of the malaria-related deaths. Although a lot of efforts have been made aiming to eradicate malaria, the World Health Organization (WHO) strives to reduce the mortality rates and malaria cases in 90% up to the year of 2030 [11]. The acquired drug resistance of the parasite continues to be a struggle in the fight against the disease, which led to rising of malaria-related death.

The resistance to antimalarial drugs is due to the indiscriminate use of the drugs and its incorrect use in treatment of malaria cases, such as wrong dosage, drug quality problems, erroneous diagnosis, not sticking to treatment, and others. These are characterized as treatment failure but can lead to a strong selective pressure in parasites, resulting in drug resistance. In recent years, with the emergence of artemisinin derivatives, resistance has allowed the number of cases to grow fast, especially in East Asia. Artemisinin, a sesquiterpene lactone, and its derivatives were adopted in the early 2000s as a first-line treatment in combined therapy for *P. falciparum* [12]. Artemisinin and its derivate can clear early trophozoites (ring stage), but the drug has a short span in vertebrate organisms, making it necessary to combined it with other drugs. In countries where *P. vivax* is the main malaria transmitter, the first-line treatment remains using chloroquine and primaquine, although WHO suggests that changes must be made to artemisinin-based combined therapy (ACT) when the rate of chloroquine resistance have reached more than 10%. Some strategies have been adapted to control malaria, such as vector control; insecticide-treated bed nets, indoor residual spraying, preventive treatment for pregnant women, and rapid diagnosis and treatment of infected individuals [1, 12, 13]. But considering the fast acquirement of resistance by parasites and vectors to drugs and insecticides, respectively, the development of an effective vaccine turns out to be an important issue. However *Plasmodium* species, especially *P. falciparum*, has a highly variant antigen pool, responsible for the adhesion of infected red blood cells (RBCs) to small vessels, which causes aggregation that leads to severe stage of the disease, making difficult advances in this area of interest [9]. The pathogenesis of *P. falciparum* relies on a complex interaction of RBC alterations, microcirculatory anomalies, and immune response. The infected RBCs start to agglomerate in small vessels by action of adhesins expressed by the parasite on the surface of infected RBCs. Those adhesins are capable of interacting with endothelial cells of small vessels, to avoid the clearance of infected RBCs by the spleen, leading to a sequestration in diverse organs, such as brain, lungs, and placenta. This, together with other factors, causes the severe forms of malaria [9].

The increasing resistance of the parasite to practically all current medications, such as artemisinin in five countries in Asia, Southeast Asia and probably South America [1], calls for the use of combination drug therapy, as well as for the identification of new targets [12, 13]. Targets targeting the parasite for the development of new therapies for the treatment of malaria encompass both cellular functions, such as detoxification of heme or ferriprotoporphyrin IX (Fe (III) PPIX), and folate metabolism, already explored for drugs established as antimalarial, as well as other metabolic pathways, such as fatty acid synthesis, and isoprenoid biosynthesis, both of which are found in the apicoplast [14].

The apicoplast, an organelle originating from a secondary endosymbiotic origin of red algae, has lost its photosynthetic function in the course of evolution [15], and speculations have demonstrated its importance in the formation of essential components incorporated into the membrane of the parasitophorous vacuole [16]. Recently, it has been shown that isoprenoid biosynthesis is not only essential for the parasite but, in fact, is the only function of the apicoplast during blood stage growth [17] and sexual forms [18]. Parasites that lacked apicoplast can be chemically rescued by addition of isopentenyl pyrophosphate (IPP) to the growth media [17].

2. Isoprenoids in *Plasmodium* spp

All isoprenoids are derived from a common precursor, IPP and its dimethylallyl pyrophosphate isomer (DMAPP) [19] (**Figure 2**). The identification and characterization of farnesyl pyrophosphate (FPP) in *P. falciparum* [20], as well as the presence of proteins covalently modified by isoprenoids [21, 22] and dolichols [23], were the first evidence for the study of isoprenoid biosynthesis in *Plasmodium*. In the last decade, there has been a broad characterization of isoprenoid biosynthesis products in the parasite [22–27] resulting from the alternative route 2-C-methyl-*d*-erythritol-4-phosphate (MEP) [28, 29] (**Figure 2**). The essential and important step in the metabolism of the biosynthesis of all isoprenoids is the elongation of the isoprene chain by enzymes called prenyltransferases. These enzymes are classified according to the chain length of the final product and the stereochemistry of the double bond formed by condensations, with FPPS (farnesyl pyrophosphate synthase) and GGPPS (geranylgeranyl pyrophosphate synthase) being the most studied prenyltransferases [30]. FPPS catalyzes the condensation of IPP with DMAPP and geranyl pyrophosphate (GPP) to form the 15-carbon isoprenoid compound, farnesyl pyrophosphate (FPP). FPP is the substrate that catalyzes the first step in the biosynthesis of ubiquinone, carotenoids, dolichols, and protein prenylation. FPP can also be condensed with an additional molecule of IPP by the enzyme GGPPS to form the 20-carbon isoprenoid, geranylgeranyl pyrophosphate (GGPP), also essential in protein isoprenylation [30] (**Figure 2**).

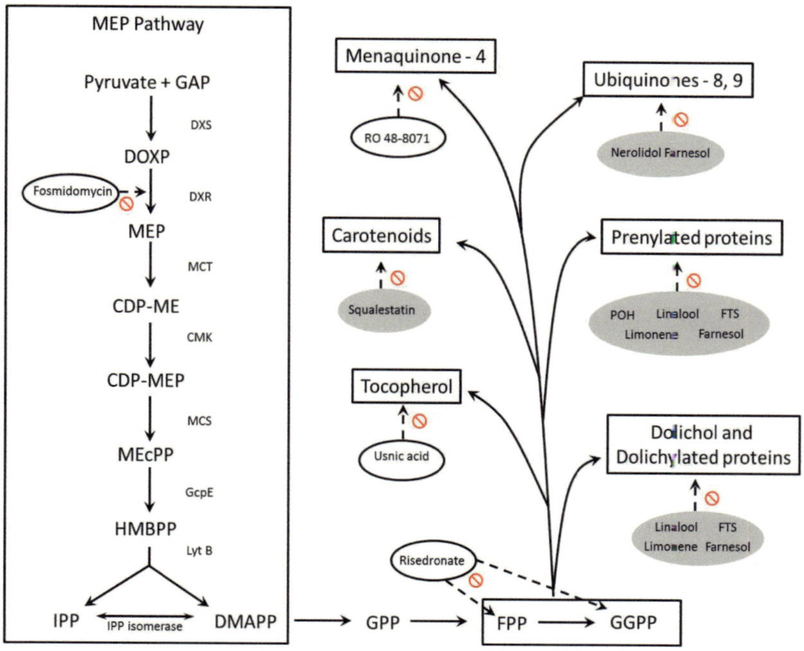

Figure 2. MEP pathway and final products in *P. falciparum*. In the left box, the MEP pathway to the formation of isoprenic units (IPP and DMAPP). In the right, final isoprenic products biosynthetized by *P. falciparum*. The circles are representing inhibitors tested against the parasite and the gray circles correspond to terpenes which its inhibitory effect on final isoprenic products biosynthesis was demonstrated. Abbreviations: *Precursors*: GAP, glyceraldehyde 3-phosphate; DOXP, 1-deoxy-*d*-xylulose 5-phosphate; MEP, 2C-methyl-*d*-erythritol 4-phosphate; CDP-ME, 4-(cytidine-5′-diphospho)-2C-methyl-*d*-erythritol; CDP-MEP, 4-(cytidine-5′-diphospho)-2C-methyl-*d*-erythritol 2-phosphate; MEcPP, 2C-methyl-*d*-erythritol 2,4-cyclodiphosphate; HMBPP, 1-hydroxy-2-methyl-2-(E)-butenyl 4-diphosphate; IPP, isopentenyl pyrophosphate; DMAPP, dimethylallyl pyrophosphate; GPP, geranyl pyrophosphate; FPP, farnesyl pyrophosphate; GGPP, geranylgeranyl pyrophosphate. *Enzymes*: DXS, 1-deoxy-*d*-xylulose 5-phosphate synthase; DXR, 1-deoxy-*d*-xylulose 5-phosphate reductoisomerase; MCT, 2C-methyl-*d*-erythritol 4-phosphate cytidine transferase; CMK, 4-(cytidine-5′-diphosph)-2C-methyl-*d*-erythritol kinase; MCS, 2C-methyl-*d*-erythritol 2,4-cyclodiphosphate synthase; GcpE, hydromethylbutenyl pyrophosphate synthase; Lyt B, hydromethylbutenyl pyrophosphate reductase; IPP isomerase, isopentenyl pyrophosphate isomerase. *Inhibitors*: FTS, s-farnesylthiosalicylic acid; POH, perillyl alcohol.

For several decades, the mevalonate pathway, present in animals and plants, was considered the only route for synthesis of the isoprene units in isoprenoid biosynthesis. The existence of a second pathway for the biosynthesis of isoprene units was discovered in 1988 by Flesch and Rohmer when they were studying the hopanoids biosynthesis (pentacyclic triterpenic steroids) in bacteria [31]. Originally called the Rohmer's pathway or mevalonate-independent pathway, its name was changed after the identification of the first step in the pathway (pyruvate/glyceraldehyde-3-phosphate GAP pathway) or the first intermediate, 1-deoxy-*d*-phosphate (DOXP pathway). However, the most accepted name is MEP [32], since this compound is the first exclusive pathway precursor. One particularly intriguing target from the MEP pathway is 1-deoxy-D-xylulose-5-phosphate synthase (DXS). DXS catalyzes the first and rate-determining step of MEP and the condensation of pyruvate and *d*-glyceraldehyde-3-phosphate to 1-deoxy-*d*-xylulose-5-phosphate (DXP) and CO_2, this reaction being the first and rate-determining step of MEP [33]. The reactions catalyzed by DXS and DXR (1-deoxy-*d*-xylulose-5-phosphate reductoisomerase) practically share the full control of the flux through MEP pathway. It was shown that reducing the flux through the pathway by inhibiting enzymes (DXS & DXR inhibition) is an important mechanism of action of some drugs leading to killing of the malarial parasite [34].

Fosmidomycin is a natural product antibiotic with activity against a number of important pathogens (**Figure 2**). It is a phosphoric acid that is a substrate mimic and direct inhibitor of the first dedicated MEP pathway enzyme, in which DXP is converted to MEP by DXR (also called IspC) [35, 36]. The use of fosmidomycin as a single-drug treatment for *P. falciparum* malaria has been hampered by low bioavailability, recrudescent, and rapid clearance from the parasite, although the compound has been used more successfully in combination with clindamycin [37]. Many efforts, with great results, have been made to improve the efficacy of fosmidomycin as modifications to the phosphonate group, extensions to the hydroxamic acid group [38, 39], and substitution of the α-position [40, 41] or, more recently, β-position [42]. The inhibition of downstream enzyme IspD (2-C-methyl-*d*-erythritol 4-phosphate cytidylyltransferase) which catalyzes the cytidylation of MEP to cytidine diphosphate methylerythritol (CDP-ME) is also metabolically apparent in fosmidomycin-treated cells. Although IspD homologs are not directly inhibited by fosmidomycin in vitro, this enzyme has been shown as a promising target in isoprenoid pathway for some new antimalarial drugs studied [43, 44].

The first evidence for the study of isoprenoid biosynthesis in *Plasmodium* were the identification and characterization of farnesyl pyrophosphate (FPP) [45], dolichols [23], and prenylated proteins in *P. falciparum* [21, 22]. Other products were functionally characterized, shown to be essential for the survival of the parasite, such as vitamin K2 (menaquinone-4) [24] and vitamin E (tocopherol) [25], for example. It was demonstrated that the vitamin K2 biosynthesized by the parasite acts as electron transporter in the respiratory chain in microaerophilic conditions and that this biosynthesis can be inhibited treating parasite culture with RO 48-8071, a specific inhibitor of manequinone-4 biosynthesis [24]. The biosynthesis of tocopherol (vitamin E) was characterized in the parasite and has been shown essential for antioxidant that protects against environmental stress, including maintaining ROS levels [46]. The parasite lives in a pro-oxidant environment that contains oxygen and iron and therefore have evolved extensive detoxifying and protective mechanisms, which both limit the production of and potential damage by ROS [27, 46]. Further, tocopherol has been showing excellent results in antimalarial drug studies, presented excellent results with usnic acid, a drug that inhibits hydroxyphenylpyruvate dioxygenase, the enzyme that catalyzes the conversion of p-hydroxyphenylpyruvic acid to homogentisic acid, a precursor of vitamin E biosynthesis [27] (**Figure 2**).

The enzymes are important tool for study the parasite physiology and an important target for antimalarial drugs, due to its specificity. The essential step in all isoprenoids biosynthesis is the elongation of the isoprene chain by enzymes called prenyltransferases. These enzymes are classified according to the chain length of the final product and the stereochemistry of the double bond formed by condensations [47]. The most studied of these prenyltransferases has been FPPS and GGPPS in *P. falciparum* [20]. The gene coding for FPPS and GGPPS have already been identified and characterized in several species such as *Saccharomyces cerevisiae* [48], *Trypanosoma cruzi* [49], *T. brucei* [50], *Toxoplasma gondii* [51], *P. vivax* [52] *and P. falciparum* [20]. Both *P. falciparum* and *T. gondii* enzymes are bifunctional (FPPS/GGPPS), being able to catalyze the biosynthesis of the isoprene compounds FPP and GGPP [20, 51], the main precursors of all secondary products from isoprenoids pathways. Transcriptional analyses of the FPPS/GGPPS gene have shown a high variability in alternative splicing of this *P. falciparum* gene [53]. It has been proven that the importance of this enzyme for the MEP pathway, due to the complexity of the gene regulation, is necessary for the formation of the main precursors. Therefore, FPPS/GGPPS is a potential and promising target for new antimalarial drugs.

Risedronate, a bisphosphonate containing nitrogen (N-BP), showed potent activity against the blood phases of *P. falciparum "in vitro"* which inhibits FPPS/GGPPS activity by competitive inhibitor toward GPP and FPP [20] (**Figure 2**). Bisphosphonates are inhibitors of bone resorption and are in clinical use for the treatment and prevention of osteoporosis [54]. The activity of the risedronate was confirmed because only farnesyl-PP and geranylgeranyl-PP restored the growth intraerythrocytic stages of *P. falciparum*, after treatment with risedronate [55]. This drug also showed a significant inhibitory effect against murine blood stage malaria, without showing toxicity effects to the animals [55] and showed great results in synergism experiments with other drugs in *P. falciparum* [56]. Also, crystallography assays of the *P. vivax* GGPP enzyme inferred that GGPP could be a major target for the lipophilic bisphosphonates [57].

Many studies of novel antimalarial drugs have been performed using enzymes as the primary target of action [58–60], although in isoprenoid biosynthesis of *P. falciparum,* few enzymes are still known and studied. The enzyme OPP/PSY (octaprenyl pyrophosphate synthase/phytoene synthase) was characterized in the parasite as a bifunctional [25, 61] responsible for form two important secondary products of the isoprenoid biosynthesis— carotenoids [25], important for antioxidant protection, and ubiquinone [24], essential for electron transfer in respiratory chain among other functions. Squalestatin, a carboxylic acid inhibitor of squalene synthase, the enzyme responsible for the first step of sterol biosynthesis, presented promising results as an antimalarial drug, specifically inhibiting OPP/PSY [62]. This results have proven then that the first carotenoid phytoene is essential for parasite development during the intraerythrocytic cycle [62], probably due its antioxidant protection function [25]. Also, through studies *in silico* characterized the secondary and tertiary structures of this enzyme, OPPS/PSY presented an unconserved unique loop in *P. falciparum* then be exploited for structure based drug designing against malaria parasite [63].

Since it was shown that the MEP pathway provides IPP precursors for the biosynthesis of higher isoprenic compounds, one of the strategies to identify secondary products of the MEP pathway was the metabolic labeling using a radioactive precursor and a posterior analysis by an appropriate method. In this context, it was identified in *P. falciparum* prenylated proteins [22], dolichols [23], ubiquinones [24], carotenoids [25], vitamin K2 [26], and vitamin E [27].

Different types of terpenes that exert antifungal, antibacterial, and antimalarial activity can be easily found in literature [19, 64, 65]. However, not all authors described a clear explanation about their mechanism of action. Recently, Silva et al. [66] listed 114 terpenes and their semisynthetic derivate with antimalarial activity, but only three have their mechanism of action elucidated. For several years, studies have been made of the large diversity of prenylated compounds biosynthesized by *P. falciparum*. That is why some research groups investigate the possibility of developing new antimalarial drugs that could interfere with the biosynthesis of isoprenoid compounds [19]. This interference in biosynthesis of isoprenoid compounds could be produced by inhibiting the enzymes as isoprenyl diphosphate synthases and isoprenyl transferases. Thus, research in the area can benefit from obtaining better knowledge about the antimalarial activity of natural terpenes. Due to their structural similarity, it was suggested that terpenes might interfere with the parasite's polyisoprenoid biosynthesis through inhibition of the isoprenyl diphosphate synthases which condense molecules of IPP among other isoprenic substrates to form isoprene chains [67]. Terpenes could also establish a competition with several enzymes which use isoprenic compounds as substrates. In fact, several terpenes, such as farnesol or linalool, have already demonstrated a capacity to inhibit at least one point of isoprenoid biosynthesis including isoprenic chain of ubiquinone and dolichol biosynthesis and protein isoprenylation [67]. These results suggest a widespread inhibitory action. Studies on antimalarial activity, drug combination, and mechanisms of action of different compounds are commonly performed *in vitro* (**Figure 2**).

It is known that IPP, FPP, and GGPP are substrates of the enzymes prenyltransferases involved in the biosynthesis of dolichol, the isoprenic side chain of ubiquinones, and the isoprenic chains attached to proteins, among other plasmodia prenylated compounds [68]. In order to determine if different drugs produce biosynthesis inhibitory effects on specific isoprenic compounds, treated cultures (using drugs concentration under the IC_{50} value), or untreated cultures can be radiolabeled by isoprenic precursors such as [1-(*n*)-^{3}H]geranylgeranyl pyrophosphate triammonium salt ([^{3}H]GGPP), 1-(*n*)-^{3}H]farnesyl pyrophosphate triammonium salt ([^{3}H]FPP), or [1-^{14}C] isopentenyl pyrophosphate triammonium salt ([^{14}C]IPP). Using these precursors, it is possible to label most isoprenic compounds without performing prenyl diphosphate pool substrate depletion, as it is necessary in other cell systems [69]. This fact suggests that the *Plasmodium* may have a different kind of isoprenic precursor uptake from the extracellular medium in comparison to other organisms [24, 67, 69]. After purifying infected red blood cells containing parasites at different stages, prenylated compounds can be isolated by diverse methods, once their radioactivity incorporation is evaluated [19]. For apolar compounds, such as dolichol and ubiquinones, the apolar extracts are commonly obtained and chromatographed by different TLC or HPLC methods, while isoprenylated proteins can easily be isolated in a polyacrylamide electrophoresis gel. Specific proteins such as Ras or Rap proteins can be easily immunoprecipitated [67, 70].

Prenylated proteins are post-translational modified by farnesyl transferase and geranylgeranyl transferase by attaching isoprenic chains to C-terminal cysteine groups. Protein prenylation had already been characterized in several parasites such as *Giardia lamblia,* [71] *Trypanosoma brucei* [72] and *Schistosoma mansoni* [73]. *P. falciparum* expresses different classes of lipidic-membrane-associated prenylated proteins such as GTPases Ras, Rho, and Rab including endosomal vesicles Rab7 protein [74] farnesylated PfPRL tyrosine phosphatase, SNARE protein Ykt6.1 [75, 76], and some putative plasmodial isoprenylated proteins such as Rab2 and Rab11a [77]. In fact, proteins of 21 to 24 kDa and 50 kDa are identified using specific prenylated protein antibodies [22, 67, 78]. Some terpenes and prenyltransferase inhibitors have already shown interesting anti-cancer and anti-plasmodial activities [21, 79]. Limonene, linalool, and perillyl alcohol (POH) are some examples of these antineoplastic and antimalarial activities, and the mechanisms of action of most of them seem to be related to protein isoprenylation interferences [67, 80–82]. Several terpenes and protein isoprenylation inhibitors were studied for its antimalarial activity. Limonene, nerolidol, farnesol, perillyl alcohol, linalool and an terpene modified as s-farnesylthiosalicylic acid (FTS) are just a few examples of compounds tested in

P. falciparum and for the IC_{50} value in 3D7 *P. falciparum* isolate was calculated [67, 82]. All these compounds have demonstrated to produce a dose–response inhibition. The lowest IC_{50} values were found for nerolidol (IC_{50}~ 760 nM), perillyl alcohol (IC_{50}~ 489 μM), FTS (IC_{50}~ 14 μM) and farnesol (IC_{50}~ 64 μM), whereas limonene and linalool showed IC_{50} values in the millimolar scale [67, 82]. All this compounds showed effects on protein isoprenylation which vary in function of the parasitic stage and the given isoprenic radiolabeled precursor. Different drug effects in function of the parasitic stage and the given isoprenic radiolabeled precursor allow us to understand compound mechanism of action at the different intraerythrocytic stages of the parasite. Treatments on protein isoprenylation seem to be specific, since there was no significant effect on L-[^{35}S] methionine-labeled proteins [22, 67].

Anti-p21*ras* or anti-p21*rap* antibodies were used for immunoprecipitation with [^{3}H]FPP or [^{3}H]GGPP labeled proteins of untreated and treated parasites at different stages. Results indicate that schizont p21*rap* radioactive incorporation is reduced after treatment with nerolidol, farnesol, linalool, limonene, and FTS. Farnesol inhibits isoprenylation of Ras and Rap proteins in all plasmodium intraerythrocytic stages whereas linalool inhibits Ras isoprenylation in ring and schizont stages. Also in the schizont stages, this effect on Ras proteins was observed as well as after FTS and limonene treatments. At ring stages, the limonene reduced the radioactive incorporation into Rap and Ras proteins, and linalool reduced incorporation into 21 kDa band. In trophozoite stages, these effects in Rap radioactive incorporation were observed after limonene and nerolidol treatments [67]. FTS is a prenylated compound that also restrains the development of some tumors by inhibiting specifically Ras post-translational modifications [83]. As mentioned, the inhibition of Ras post-translational modifications in schizont stage and the hindrance of the parasite at this stage were also demonstrated for *P. falciparum*.

Carotenoids, dolichol, and coenzyme Q are examples of prenylated compounds biosynthesized by *P. falciparum* isolated by diverse methods. Coenzyme Q has a role in mitochondrial respiration, in pyrimidine biosynthesis, and in preventing membrane lipoperoxidation, among other functions [84]. Carotenoids are believed to avoid oxidative stress; and dolichol and its phosphorylated derivatives can participate in protein prenylation [85] or act as lipids' carriers, being used for several glycoconjugates [86] including N-linked glycoproteins. Those play an important role in the parasite's intraerythrocytic stages differentiation [87]. The *P. falciparum* bifunctional enzyme OPPS/PSY is involved in the biosynthesis of carotenoids and isoprenic side chains attached to benzoquinone rings derived from *p*-hydroxybenzoic acid. The parasite is able to synthesize several carotenoids and at least two homologs of coenzyme Q (Q_8 and Q_9) that differ on the isoprenic units of the lateral chain [24, 25, 61]. Some authors suggested that the homologs might act against different kinds of oxidative stress conditions [24]. Several terpenes also demonstrated effects on dolichol and ubiquinone biosynthesis in different organisms. In hepatoma and Neuro2A cells, the monoterpenes limonene and linalool interfered in the biosynthesis of dolichol and the isoprenic side chains of benzoquinones [81]. Similarly, in *P. falciparum*, both monoterpenes and also farnesol and nerolidol interfered in the biosynthesis of dolichol and ubiquinones when the parasite was [^{14}C]IPP labeled. Only the monoterpenes limonene and linalool interfered in geraniol and farnesol radioactive incorporation. That is why some authors suggested that limonene and linalool could be interfering in the condensation between IPP and DMAPP, while farnesol or nerolidol could be interfering in the isoprenic chains' elongation [67]. In order to demonstrate terpenes, specifically effects on

P. falciparum isoprenoid-related enzymes, nerolidol was used to inhibit the OPPS/PSY activity using [^{3}H]FPP and [^{14}C]IPP as substrates. OPPS/PSY activity was competitively inhibited by nerolidol (K_i 15 nM) [61].

As we have seen, some MEP pathway and isoprenoid pathway inhibitors show good antimalarial activity and produce important metabolic alterations in *P. falciparum*. Different authors asked if using terpenes individually or in combination with other drugs could be a good strategy to treat malaria. It is well established that fosmidomycin and clindamycin, when combined, produce a synergistic activity *in vitro* and *in vivo* [88], and several drugs which act at different points of the MEP/isoprenoid showed supra-additive *in vitro* effects when combined [56]. These kinds of drug combination studies are useful for a better understanding of the interactions between the different intermediates of the MEP and isoprenoid pathways and to evaluate its antimalarial potential. Fosmidomycin, risedronate, nerolidol and squalestatin (supposed to be a phytoene synthase inhibitor) are drugs which are believed to act in different cellular compartments, and all of them seem to inhibit at least one point related to isoprenoid metabolism [56]. Except nerolidol-risedronate, most binary combinations between fosmidomycin, risedronate, nerolidol and squalestatin showed a supra-additive effect [56]. Probably it is because they target different enzymes in the same biosynthetic pathway. On the other hand, nerolidol and risedronate when combined showed a sub-additive effect. It was suggested that this fact could be explained because both compounds affect the same target of isoprenoid pathways [56]. Nerolidol affects the synthesis of several isoprenoid compounds, including protein isoprenylation as well as risedronate [55, 67].

Due to the terpenes effectiveness to inhibit the *P. falciparum* growth *in vitro*, several studies have been focused on studying these effects on experimental models of malaria by *P. berghei* infection. Nerolidol for example, being administered on Balb/c mice at a dose 2000 mg/kg/day by oral and intranasal via, had an inhibitory effect on the *P. berghei* ANKA growth, since the parasitemia were reduced and the survival rates were increased significantly with the nerolidol treatment [89]. Also, some derivatives of 4-nerolidylcatechol, at oral doses of 50 mg/kg/day, had suppressed *P. berghei* NK65 in infected BALB/c mice by 44%, showing marked improvement over the parasite's growth [90].

On the other hand, some metabolites derived from limonene, such as perillyl alcohol, have also been shown to be effective against the severe conditions development caused by *P. berghei* infection. C57BL/6 mice infected with *P. berghei* ANKA and treated with 500 mg/kg/day intranasal via had a significant increase in survival rates, showing a preventive effect against the experimental cerebral malaria development [82].

Furthermore, plant extracts that contain several terpenes have also been tested on experimental models. The antimalarial activity of ethanolic bark extract of *A. lebbeck* was determined. Phytochemical screening revealed the presence of alkaloids, flavonoids, phenols, saponins, terpenes, and phytosterols [91].

Limonoids isolated from the residual seed biomass from *Carapa guianensis* were tested against *P. berghei* and 6α-acetoxy-gedunin was more active than 7-deacetoxy-7-oxogedunin. At oral doses of 50 and 100 mg/kg/day, 6α-acetoxy-gedunin suppressed parasitemia versus untreated controls by 40 and 66%, respectively, evidencing a clear dose response [92].

In other studies, methanol extracts of *Carpesium rosulatum* were found to have potential antimalarial activity *in vivo* when tested against *P. berghei* in mice. A dose of 2, 5, 10 mg/kg/day exhibited a significant blood schizonticidal activity in four-day early infection with a significant mean survival time comparable to that of the standard drug, chloroquine (5 mg/kg/day) [93].

Some others triterpenes isolated from the African medicinal plant, *Momordica balsamina* L. such as, balsaminoside B, karavilagenin C, and the karavoates B and D were synthesized by diacylation from these extracted terpenes. Derivatives exhibited sub-micromolar IC_{50} *in vitro* against *P. falciparum* strains and exhibited greater *in vivo* antimalarial activity. Orally and subcutaneously administered karavoate B exhibited the greatest *in vivo* antimalarial activity (55.2–58.1% maximal suppression of parasitemia at doses of 50 mg/kg/day) [94].

3. Conclusion

Plasmodium spp. has an organelle, the apicoplast, which is essential for the development of the parasite because it is linked to two metabolic pathways, one of which is the isoprenoid biosynthesis that is different in several steps from the isoprenoid pathway in the vertebrate host. The biosynthesis pathway is an important target for evaluating new antimalarial drugs and the terpenes for being derived from the isoprenoid pathway and having a similar structure can interfere in the synthesis of isoprenoids and should be evaluated as antimalarial potentials.

Acknowledgements

This work was supported by grants from Brazilian Agencies CNPq and FAPESP (Brazil). H.B.G., R.A.C.S., A.A.M.R., I.B.V., and G.C.F.L. are fellowships from FAPESP.

Author details

Heloisa Berti Gabriel, Rodrigo AC Sussmann, Emila A Kimura,
Adriana Alejandra Marin Rodriguez, Ignasi Bofill Verdaguer,
Gabriela Carolina Fernandes Leite and Alejandro Miguel Katzin*

*Address all correspondence to: amkatzin@icb.usp.br

Department of Parasitology, Institute of Biomedical Sciences, University of São Paulo,
São Paulo, Brazil

References

[1] World Health Organization. World Malaria Report 2017. Geneva. 217

[2] Grigg MJ, Snounou G. Plasmodium simium: A Brazilian focus of anthropozoonotic vivax malaria? The Lancet Global Health. 2017;**5**:e961-e962. DOI: 10.1016/S2214-109X(17)30343-1

[3] Camargo EP. Malária, maleita, paludismo. Ciência e Cultura. 2003;**55**:26-29

[4] Burda P-C, Caldelari R, Heussler VT. Manipulation of the host cell membrane during plasmodium liver stage egress. MBio. 2017;**8**:e00139-e00117. DOI: 10.1128/mBio.00139-17

[5] Miller LH, Ackerman HC, Su X, Wellems TE. Malaria biology and disease pathogenesis: Insights for new treatments. Nature Medicine. 2013;**19**:156-167. DOI: 10.1038/nm.3073

[6] Cowman AF, Berry D, Baum J. The cellular and molecular basis for malaria parasite invasion of the human red blood cell. The Journal of Cell Biology. 2012;**198**:961-971. DOI: 10.1083/jcb.201206112

[7] Coatney GR. The simian malarias: Zoonoses, anthroponoses, or both? The American Journal of Tropical Medicine and Hygiene. 1971;**20**:795-803

[8] Martinelli A, Cullenton R. Non-human primate malaria parasites: Out of the forest and into the laboratory. Parasitology. 2018:**145**(1):41-54. DOI: 10.1017/S0031182016001335

[9] Miller LH, Baruch DI, Marsh K, Doumbo OK. The pathogenic basis of malaria. Nature. 2002;**415**:673-679. DOI: 10.1038/415673a

[10] Meibalan E, Marti M. Biology of malaria transmission. Cold Spring Harbor Perspectives in Medicine. 2017;**7**:a025452. DOI: 10.1101/cshperspect.a025452

[11] World Health Organization. Global Technical Strategy for Malaria 2016-2030. Geneva: WHO Press; 2015

[12] Blasco B, Leroy D, Fidock DA. Antimalarial drug resistance: Linking plasmodium falciparum parasite biology to the clinic. Nature Medicine. 2017;**23**:917-928. DOI: 10.1038/nm.4381

[13] Burrows JN, Hooft van Huijsduijnen R, Möhrle JJ, Oeuvray C, Wells TN. Designing the next generation of medicines for malaria control and eradication. Malaria Journal. 2013;**12**:187. DOI: 10.1186/1475-2875-12-187

[14] Ginsburg H. The biochemistry of *Plasmodium falciparum*. In: Advances in Malaria Research. Hoboken, NJ, USA, John Wiley & Sons, Inc.; 2016. pp. 219-290. DOI: 10.1002/9781118493816.ch9

[15] Arisue N, Hashimoto T. Phylogeny and evolution of apicoplasts and apicomplexan parasites. Parasitology International. 2015;**64**:254-259. DOI: 10.1016/j.parint.2014.10.005

[16] Wilson RJM. Progress with parasite plastids. Journal of Molecular Biology. 2002;**319**:257-274. DOI: 10.1016/S0022-2836(02)00303-0

[17] Yeh E, DeRisi JL. Chemical rescue of malaria parasites lacking an Apicoplast defines organelle function in blood-stage plasmodium falciparum. PLoS Biology. 2011;**9**:e1001138. DOI: 10.1371/journal.pbio.1001138

[18] Wiley JD, Merino EF, Krai PM, McLean KJ, Tripathi AK, Vega-Rodríguez J, et al. Isoprenoid precursor biosynthesis is the essential metabolic role of the Apicoplast during Gametocytogenesis in plasmodium falciparum. Eukaryotic Cell. 2015;**14**:128-139. DOI: 10.1128/EC.00198-14

[19] Jordão FM, Kimura EA, Katzin AM. Isoprenoid biosynthesis in the erythrocytic stages of plasmodium falciparum. Memórias do Instituto Oswaldo Cruz. 2011;**106**:134-141. DOI: 10.1590/S0074-02762011000900018

[20] Jordão FM, Gabriel HB, Alves JM, Angeli CB, Bifano TD, Breda A, et al. Cloning and characterization of bifunctional enzyme farnesyl diphosphate/geranylgeranyl diphosphate synthase from plasmodium falciparum. Malaria Journal. 2013;**12**:184. DOI: 10.1186/1475-2875-12-184

[21] Chakrabarti D, Da Silva T, Barger J, Paquette S, Patel H, Patterson S, et al. Protein Farnesyltransferase and protein Prenylation in plasmodium falciparum. The Journal of Biological Chemistry. 2002;**277**:42066-42073. DOI: 10.1074/jbc.M202860200

[22] Moura IC, Wunderlich G, Uhrig ML, Couto AS, Kimura LIAA. Peres VJ, et al. Limonene Arrests Parasite Development and Inhibits Isoprenylation of Proteins in *Plasmodium falciparum*. 2001;**45**:2553-2558. DOI: 10.1128/AAC.45.9.2553

[23] Couto AS, Kimura EA, Peres VJ, Uhrig ML, Katzin AM. *Plasmodium falciparum*: Presence of Dolichols of 11 and 12 Isoprene Units. 1999;**637**:629-637

[24] Macedo CS, Uhrig ML, Kimura EA, Katzin AM. Characterization of the isoprenoid chain of coenzyme Q in plasmodium falciparum. FEMS Microbiology Letters. 2002;**207**:13-20. DOI: 10.1111/j.1574-6968.2002.tb11021.x

[25] Tonhosolo R, D'Alexandri FL, de Rosso VV, Gazarini ML, Matsumura MY, Peres VJ, et al. Carotenoid biosynthesis in Intraerythrocytic stages of plasmodium falciparum. The Journal of Biological Chemistry. 2009;**284**:9974-9985. DOI: 10.1074/jbc.M807464200

[26] Tonhosolo R, Gabriel HB, Matsumura MY, Cabral FJ, Yamamoto MM, D'Alexandri FL, et al. Intraerythrocytic stages of plasmodium falciparum biosynthesize menaquinone. FEBS Letters. 2010;**584**. DOI: 10.1016/j.febslet.2010.10.055

[27] Sussmann RAC, Angeli CB, Peres VJ, Kimura EA, Katzin AM. Intraerythrocytic stages of plasmodium falciparum biosynthesize vitamin E. FEBS Letters. 2011;**585**:3985-3991. DOI: 10.1016/j.febslet.2011.11.005

[28] Jomaa H, Wiesner J, Sanderbrand S, Altincicek B, Weidemeyer C, Hintz M, et al. Inhibitors of the nonmevalonate pathway of isoprenoid biosynthesis as antimalarial drugs. Science. 1999;**285**:1573-1576. DOI: 10.1126/science.285.5433.1573

[29] Cassera MB, Gozzo FC, D'Alexandri FL, Merino EF, del Portillo HA, Peres VJ, et al. the Methylerythritol phosphate pathway is functionally active in all Intraerythrocytic stages of *Plasmodium falciparum*. The Journal of Biological Chemistry. 2004;**279**:51749-51759. DOI: 10.1074/jbc.M408360200

[30] Ogura K, Koyama T. Enzymatic aspects of Isoprenoid chain elongation. Chemical Reviews. 1998;**98**:1263-1276

[31] Flesch G, Rohmer M. Prokaryotic hopanoids: The biosynthesis of the bacteriohopane skeleton. Formation of isoprenic units from two distinct acetate pools and a novel type of carbon/carbon linkage between a triterpene and D-ribose. European Journal of Biochemistry. 1988;**175**:405-411

[32] Eisenreich W, Bacher A, Arigoni D, Rohdich F. Biosynthesis of isoprenoids via the non-mevalonate pathway. Cellular and Molecular Life Sciences. 2004;**61**:1401-1426. DOI: 10.1007/s00018-004-3381-z

[33] Handa S, Ramamoorthy D, Spradling TJ, Guida WC, Adams JH, Bendinskas KG, et al. Production of recombinant 1-deoxy-d-xylulose-5-phosphate synthase from plasmodium vivax in Escherichia coli. FEBS Open Bio. 2013;**3**:124-129. DOI: 10.1016/j.fob.2013.01.007

[34] Edwards RL, Brothers RC, Wang X, Maron MI, Ziniel PD, Tsang PS, et al. MEPicides: Potent antimalarial prodrugs targeting isoprenoid biosynthesis. Scientific Reports. 2017; **7**:8400. DOI: 10.1038/s41598-017-07159-y

[35] Imlay L, Odom AR. Isoprenoid metabolism in Apicomplexan parasites. Current Clinical Microbiology Reports. 2014;**1**:37-50. DOI: 10.1007/s40588-014-0006-7

[36] Jansson AM, Więckowska A, Björkelid C, Yahiaoui S, Sooriyaarachchi S, Lindh M, et al. DXR inhibition by potent mono- and Disubstituted Fosmidomycin analogues. Journal of Medicinal Chemistry. 2013;**56**:6190-6199. DOI: 10.1021/jm4006498

[37] Borrmann S, Adegnika AA, Matsiegui P, Issifou S, Schindler A, Mawili Mboumba DP, et al. Fosmidomycin/clindamycin for plasmodium falciparum infections in African children. The Journal of Infectious Diseases. 2004;**189**:901-908. DOI: 10.1086/381785

[38] Zinglé C, Kuntz L, Tritsch D, Grosdemange-Billiard C, Rohmer M. Isoprenoid biosynthesis via the Methylerythritol phosphate pathway: Structural variations around Phosphonate anchor and spacer of Fosmidomycin, a potent inhibitor of Deoxyxylulose phosphate Reductoisomerase. The Journal of Organic Chemistry. 2010;**75**:3203-3207. DOI: 10.1021/jo9024732

[39] Umeda T, Tanaka N, Kusakabe Y, Nakanishi M, Kitade Y, Nakamura KT. Molecular basis of fosmidomycin's action on the human malaria parasite plasmodium falciparum. Scientific Reports. 2011;**1**:9. DOI: 10.1038/srep00009

[40] Kunfermann A, Lienau C, Illarionov B, Held J, Gräwert T, Behrendt CT, et al. IspC as target for Antiinfective drug discovery: Synthesis, Enantiomeric separation, and structural biology of Fosmidomycin Thia Isosters. Journal of Medicinal Chemistry. 2013;**56**:8151-8162. DOI: 10.1021/jm4012559

[41] Konzuch S, Umeda T, Held J, Hähn S, Brücher K, Lienau C, et al. Binding modes of reverse Fosmidomycin analogs toward the antimalarial target IspC. Journal of Medicinal Chemistry. 2014;**57**:8827-8838. DOI: 10.1021/jm500850y

[42] Chofor R, Sooriyaarachchi S, Risseeuw MDP, Bergfors T, Pouyez J, Johny C, et al. Synthesis and bioactivity of β-substituted Fosmidomycin analogues targeting 1-Deoxy-d-xylulose-5-phosphate Reductoisomerase. Journal of Medicinal Chemistry. 2015;**58**:2988-3001. DOI: 10.1021/jm5014264

[43] Wu W, Herrera Z, Ebert D, Baska K, Cho SH, DeRisi JL, et al. A chemical rescue screen identifies a plasmodium falciparum apicoplast inhibitor targeting MEP isoprenoid precursor biosynthesis. Antimicrobial Agents and Chemotherapy. 2015;**59**:356-364. DOI: 10.1128/AAC.03342-14

[44] Imlay LS, Armstrong CM, Masters MC, Li T, Price KE, Edwards RL, et al. Plasmodium IspD (2-C-methyl-D-erythritol 4-phosphate cytidyltransferase), an essential and druggable antimalarial target. ACS Infectious Diseases. 2015;**1**:157-167. DOI: 10.1021/id500047s

[45] Mbaya B, Rigomier D, Edorh GG, Karst F, Schrevel J. Isoprenoid metabolism in *Plasmodium falciparum* during the intraerythrocytic phase of malaria. Biochemical and Biophysical Research Communications. 1990;**173**:849-854

[46] Sussmann RAC, Fotoran WL, Kimura EA, Katzin AM. *Plasmodium falciparum* uses vitamin e to avoid oxidative stress. Parasites and Vectors. 2017;**10**:461-468 DOI: 10.1186/s13071-017-2402-3

[47] Liang P-H, Ko T-P, Wang AH-J. Structure, mechanism and function of prenyltransferases. European Journal of Biochemistry. 2002;**269**:3339-3354. DOI: 10.1046/j.1432-1033.2002.03014.x

[48] Anderson MS, Yarger JG, Burck CL, Poulter CD. Farnesyl diphosphate synthetase. Molecular cloning, sequence, and expression of an essential gene from *Saccharomyces cerevisiae*. The Journal of Biological Chemistry. 1989;**264**:19176-19184

[49] Montalvetti A, Bailey BN, Martin MB, Severin GW, Oldfield E, Docampo R. Bisphosphonates are potent inhibitors of *Trypanosoma cruzi* Farnesyl pyrophosphate synthase. The Journal of Biological Chemistry. 2001;**276**:33930-33937. DOI: 10.1074/jbc.M103950200

[50] Montalvetti A, Fernandez A, Sanders JM, Ghosh S, Van Brussel E, Oldfield E, et al. Farnesyl pyrophosphate synthase is an essential enzyme in *Trypanosoma brucei*. The Journal of Biological Chemistry. 2003;**278**:17075-17083. DOI: 10.1074/jbc.M210467200

[51] Ling Y, Li Z-H, Miranda K, Oldfield E, Moreno SNJ. The farnesyl-diphosphate/geranylgeranyl-diphosphate synthase of toxoplasma gondii is a bifunctional enzyme and a molecular target of bisphosphonates. The Journal of Biological Chemistry. 2007;**282**:30804-30816. DOI: 10.1074/jbc.M703178200

[52] Artz JD, Wernimont AK, Dunford JE, Schapira M, Dong A, Zhao Y, et al. Molecular characterization of a novel geranylgeranyl pyrophosphate synthase from plasmodium parasites. The Journal of Biological Chemistry. 2011;**286**:3315-3322. DOI: 10.1074/jbc.M109.027235

[53] Gabriel HB, de Azevedo MF, Palmisano G, Wunderlich G, Kimura EA, Katzin AM, et al. Single-target high-throughput transcription analyses reveal high levels of alternative

splicing present in the FPPS/GGPPS from *Plasmodium falciparum*. Scientific Reports. 2016;**5**:18429. DOI: 10.1038/srep18429

[54] Coxon FP, Helfrich MH, Van't Hof R, Sebti S, Ralston SH, Hamilton A, et al. Protein Geranylgeranylation is required for osteoclast formation, function, and survival: Inhibition by bisphosphonates and GGTI-298. Journal of Bone and Mineral Research. 2000;**15**:1467-1476. DOI: 10.1359/jbmr.2000.15.8.1467

[55] Jordao FM, Saito AY, Miguel DC, de Jesus PV, Kimura EA, Katzin AM. In vitro and in vivo antiplasmodial activities of risedronate and its interference with protein prenylation in *Plasmodium falciparum*. Antimicrobial Agents and Chemotherapy. 2011;**55**:2026-2031. DOI: 10.1128/AAC.01820-10

[56] da Silva MF, Saito AY, Peres VJ, Oliveira AC, Katzin AM. *In Vitro* antimalarial activity of different inhibitors of the plasmodial isoprenoid synthesis pathway. Antimicrobial Agents and Chemotherapy. 2015;**59**:5084-5087. DOI: 10.1128/AAC.04161-14

[57] No JH, de Macedo DF, Zhang Y, Liu Y-L, Zhu W, Feng X, et al. Lipophilic analogs of zoledronate and risedronate inhibit plasmodium geranylgeranyl diphosphate synthase (GGPPS) and exhibit potent antimalarial activity. Proceedings of the National Academy of Sciences. 2012;**109**:4058-4063. DOI: 10.1073/pnas.1118215109

[58] Kato N, Comer E, Sakata-Kato T, Sharma A, Sharma M, Maetani M, et al. Diversity-oriented synthesis yields novel multistage antimalarial inhibitors. Nature. 2016;**538**:344-349. DOI: 10.1038/nature19804

[59] Saint-Léger A, Sinadinos C, Ribas de Pouplana L. The growing pipeline of natural aminoacyl-tRNA synthetase inhibitors for malaria treatment. Bioengineered. 2016;**7**:60-64. DOI: 10.1080/21655979.2016.1149270

[60] Deu E. Proteases as antimalarial targets: Strategies for genetic, chemical, and therapeutic validation. The FEBS Journal. 2017;**284**:2604-2628. DOI: 10.1111/febs.14130

[61] Tonhosolo R, D'Alexandri FL, Genta FA, Wunderlich G, Gozzo FC, Eberlin MN, et al. Identification, molecular cloning and functional characterization of an octaprenyl pyrophosphate synthase in intra-erythrocytic stages of *Plasmodium falciparum*. Biochemical Journal 2005;392:117-126. doi:BJ20050441 [pii]10.1042/BJ20050441

[62] Gabriel HB, Silva MF, Kimura EA, Wunderlich G, Katzin AM, Azevedo MF. Squalestatin is an inhibitor of carotenoid biosynthesis in *Plasmodium falciparum*. Antimicrobial Agents and Chemotherapy. 2015;**59**:3180-3188. DOI: 10.1128/AAC.04500-14

[63] Agarwal S, Sharma V, Phulera S, Abdin MZ, Ayana R, Singh S. Structural insights into a key carotenogenesis related enzyme phytoene synthase of *P. Falciparum*: A novel drug target for malaria. Systems and Synthetic Biology. 2015;**9**:27-37. DOI: 10.1007/s11693-015-9168-8

[64] Rohmer M. The discovery of a mevalonate-independent pathway for isoprenoid biosynthesis in bacteria, algae and higher plants. Natural Product Reports. 1999;**16**:565-574. DOI: 10.1039/a709175c

[65] Cowan MM. Plant products as antimicrobial agents. Clinical Microbiology Reviews. 1999;**12**:564-582

[66] Silva GNS, Rezende LCD, Emery FS, Gosmann G, Gnoatto SCB. Natural and semi synthetic antimalarial compounds: Emphasis on the terpene class. Mini Reviews in Medicinal Chemistry. 2015;**15**:809-836

[67] Goulart HR, Kimura EA, Peres VJ, Couto AS, Duarte FAA, Katzin AM. Antimicrobial Agents and Chemotherapy. Terpenes arrest parasite development and inhibit biosynthesis of isoprenoids in *Plasmodium falciparum*. 2004;**48**:2502-2509. DOI: 10.1128/AAC.48.7.2502

[68] Wang KC, Ohnuma S. Isoprenyl diphosphate synthases. Biochimica et Biophysica Acta. 2000;**1529**:33-48

[69] Bittman R. Cholesterol: Its Functions and Metabolism in Biology and Medicine. Vol. 28. Springer US: Boston, MA; 1997. DOI: 10.1007/978-1-4615-5901-6

[70] Hand PH, Thor A, Wunderlich D, Muraro R, Caruso A, Schlom J. Monoclonal antibodies of predefined specificity detect activated ras gene expression in human mammary and colon carcinomas. Proceedings of the National Academy of Sciences of the United States of America. 1984;81:5227-5231

[71] Luján HD, Mowatt MR, Chen G-Z, Nash TE. Isoprenylation of proteins in the protozoan Giardia lamblia. Molecular and Biochemical Parasitology. 1995;**72**:121-127. DOI: 10.1016/0166-6851(94)00070-4

[72] Field H, Blench I, Croft S, Field MC. Characterisation of protein isoprenylation in procyclic form *Trypanosoma brucei*. Molecular and Biochemical Parasitology. 1996;**82**:67-80. DOI: 10.1016/0166-6851(96)02723-5

[73] Chen G-Z, Bennett JL. Characterization of mevalonate-labeled lipids isolated from parasite proteins in *Schistosoma mansoni*. Molecular and Biochemical Parasitology. 1993;**59**:287-292. DOI: 10.1016/0166-6851(93)90226-N

[74] Krai P, Dalal S, Klemba M. Evidence for a Golgi-to-endosome protein sorting pathway in *Plasmodium falciparum*. PLoS One. 2014;**9**:e89771. DOI: 10.1371/journal.pone.0089771

[75] Ayong L, DaSilva T, Mauser J, Allen CM, Chakrabarti D. Evidence for prenylation-dependent targeting of a Ykt6 SNARE in *Plasmodium falciparum*. Molecular and Biochemical Parasitology. 2011;**175**:162-168. DOI: 10.1016/j.molbiopara.2010.11.007

[76] Pendyala PR, Ayong L, Eatrides J, Schreiber M, Pham C, Chakrabarti R, et al. Characterization of a PRL protein tyrosine phosphatase from *Plasmodium falciparum*. Molecular and Biochemical Parasitology. 2008;**158**:1-10. DOI: 10.1016/j.molbiopara.2007.11.006

[77] Buckner FS, Eastman RT, Yokoyama K, Gelb MH, Van Voorhis WC. Protein farnesyl transferase inhibitors for the treatment of malaria and African trypanosomiasis. Current Opinion in Investigational Drugs. 2005;**6**:791-797

[78] Howe R, Kelly M, Jimah J, Hodge D, Odom AR. Isoprenoid biosynthesis inhibition disrupts Rab5 localization and food vacuolar integrity in *Plasmodium falciparum*. Eukaryotic Cell. 2013;**12**:215-223. DOI: 10.1128/EC.00073-12

[79] Carrico D, Ohkanda J, Kendrick H, Yokoyama K, Blaskovich MA, Bucher CJ, et al. In vitro and in vivo antimalarial activity of peptidomimetic protein farnesyltransferase inhibitors with improved membrane permeability. Bioorganic & Medicinal Chemistry. 2004;**12**:6517-6526. DOI: 10.1016/j.bmc.2004.09.020

[80] Holstein SA, Hohl RJ. Monoterpene regulation of Ras and Ras-related protein expression. Journal of Lipid Research. 2003;**44**:1209-1215. DOI: 10.1194/jlr.M300057-JLR200

[81] Kawata S, Nagase T, Yamasaki E, Ishiguro H, Matsuzawa Y. No titlemodulation of the mevalonate pathway and cell growth by pravastatin and d-limonene in a human hepatoma cell line (Hep G2). British Journal of Cancer. 1994;**69**:1015-1020

[82] Rodriguez AAM, Carvalho LJM, Kimura EA, Katzin AM. Perillyl alcohol exhibits in vitro inhibitory activity against *Plasmodium falciparum* and protects against experimental cerebral malaria. International Journal of Antimicrobial Agents. 2017;**51**(3):370-377. DOI: 10.1016/j.ijantimicag.2017.08.025

[83] Egozi Y, Weisz B, Gana-Weisz M, Ben-Baruch G, Kloog Y. Growth inhibition of ras-dependent tumors in nude mice by a potent ras-dislodging antagonist. International Journal of Cancer. 1999;**80**:911-918. DOI: 10.1002/(SICI)1097-0215(19990315)80:6<911::AID-IJC18>3.0.CO;2-4

[84] Nowicka B, Kruk J. Occurrence, biosynthesis and function of isoprenoid quinones. Biochimica et Biophysica Acta (BBA) - Bioenergetics. 1797;**2010**:1587-1605. DOI: 10.1016/j.bbabio.2010.06.007

[85] D'Alexandri FL, Kimura EA, Peres VJ, Katzin AM. Protein dolichylation in *Plasmodium falciparum*. FEBS Letters. 2006;**580**:6343-6348. DOI: 10.1016/j.febslet.2006.10.042

[86] Schwarz RT, Datema R. The lipid pathway of protein glycosylation and its inhibitors: The biological significance of protein-bound carbohydrates. Advances in Carbohydrate Chemistry and Biochemistry. 1982;**40**:287-379

[87] Kimura EA, Couto AS, Peres VJ, Casal OL, Katzin AM. N -linked glycoproteins are related to schizogony of the intraerythrocytic stage in *Plasmodium falciparum*. The Journal of Biological Chemistry. 1996;**271**:14452-14461. DOI: 10.1074/jbc.271.24.14452

[88] Wiesner J, Henschker D, Hutchinson DB, Beck E, Jomaa H. In vitro and in vivo synergy of fosmidomycin, a novel antimalarial drug, with clindamycin. Antimicrobial Agents and Chemotherapy. 2002;**46**:2889-2894. DOI: 10.1128/AAC.46.9.2889-2894.2002

[89] Saito AY, Marin Rodriguez AA, Menchaca Vega DS, Sussmann RAC, Kimura EA, Katzin AM. Antimalarial activity of the terpene nerolidol. International Journal of Antimicrobial Agents. 2016;**48**. DOI: 10.1016/j.ijantimicag.2016.08.017

[90] E Silva LFR, Nogueira KL, Da Silva Pinto AC, Katzin AM, Sussmann RAC, Muniz MP, et al. In vivo antimalarial activity and mechanisms of action of 4-nerolidylcatechol

derivatives. Antimicrobial Agents and Chemotherapy. 2015;**59**:3271-3280. DOI: 10.1128/AAC.05012-14

[91] Kalia S, Walter N, Bagai U. Antimalarial efficacy of *Albizia lebbeck* (Leguminosae) against *Plasmodium falciparum* in vitro & *P. berghei* in vivo. The Indian Journal of Medical Research. 2015;**142**:101. DOI: 10.4103/0971-5916.176635

[92] Pereira TB, Rocha e Silva LF, Amorim RC, Melo MR, Zacardi de Souza RC, Eberlin MN, et al. In vitro and in vivo anti-malarial activity of limonoids isolated from the residual seed biomass from *Carapa guianensis* (andiroba) oil production. Malaria Journal. 2014;**13**:317. DOI: 10.1186/1475-2875-13-317

[93] Chung I-M, Kim M-Y, Moon H-I. Antiplasmodial activity of sesquiterpene lactone from *Carpesium rosulatum* in mice. Parasitology Research. 2008;**103**:341-344. DOI: 10.1007/s00436-008-0977-5

[94] Rocha e Silva LF, Ramalhete C, Nogueira KL, Mulhovo S, Ferreira M-JU, Pohlit AM. In vivo evaluation of isolated triterpenes and semi-synthetic derivatives as antimalarial agents. European Journal of Medicinal Chemistry. 2015;**102**:398-402. DOI: 10.1016/j.ejmech.2015.08.022

Chapter 4

Terpenes from Natural Products with Potential Anti-Inflammatory Activity

Roberto José Serrano Vega, Nimsi Campos Xolalpa,
Angel Josabad Alonso Castro,
Cuauhtémoc Pérez González, Julia Pérez Ramos and
Salud Pérez Gutiérrez

Additional information is available at the end of the chapter

http://dx.doi.org/10.5772/intechopen.73215

Abstract

The development and progression of many diseases is related with an inflammatory process, which could affect different organs or tissues. Currently, many drugs are used to treat inflammation. However, some of these compounds induce severe side effects. For this reason, the search of new therapeutic options for the treatment of inflammation is very desirable. Medicinal plants have been an interesting source for obtaining new active compounds, including several terpenes and terpenoids with anti-inflammatory activity. This book chapter includes 62 sesquiterpenes, 34 diterpenes, and 22 triterpenes with anti-inflammatory activity. The anti-inflammatory effect was evaluated using *in vitro, in vivo,* and both models. These terpenes were obtained from 44 plant species belonging to 25 botanical families. Eight of theses species belong to the Asteraceae family and four to Lamiaceae family, respectively, and the other species belong to 13 different botanical families, one sesquiterpene was obtained from a sponge and two diterpenes were isolated from corals.

Keywords: Terpenes, terpenoids, anti-inflammatory activity, natural, products

1. Inflammation

Inflammation is a response of vascularized tissues to infections and tissue damage, and contributes to the beginning and progression of diseases such as Alzheimer, type 2 diabetes, obesity, stroke, and cancer [1, 2]. The symptomatology of inflammation is characterized by pain, redness, swelling, heat, and loss of function. Depending on the time of duration, inflammation might be categorized into acute and chronic. Acute inflammation is considered as a protective response and occurs within minutes, hours, or few days after exposure to infections and/or tissue damage. Acute inflammation is characterized by the exudation of fluid (edema), elevated blood flow, and migration of neutrophils [3]. Chronic inflammation occurs when the initial response fails to repair tissue damaged or when a noxious stimulus is persistent, and is characterized with more tissue destruction, fibrosis, long presence of lymphocytes [4]. Macrophages, dendritic cells, and mast cells initiate the inflammatory process secreting pro-inflammatory cytokines such as interleukin 6 (IL-6) and interleukin 8 (IL-8), tumor necrosis factor-α (TNF-α), and inducing the production of reactive oxygen species (ROS), which play an important role in the modulation of inflammation [5]. The long-term use of current drugs for the treatment of inflammation, including non-steroidal anti-inflammatory drugs (NSAIDs), the disease-modifying anti-rheumatic drugs (DMARDs), and steroids display several undesirable side effects such as gastric ulcers, nephrotoxicity, and hepatotoxicity, among others [6]. The search of new anti-inflammatory agents with less side effects is highly desirable. Furthermore, the efficient treatment of inflammation may be an interesting and effective way to prevent chronic diseases like cancer.

2. Terpenes

Natural products are a good source of anti-inflammatory compounds [7]. Terpenes, containing a C5 isoprene unit, are the large group of natural compounds found mostly in higher plants, but also in lower invertebrates. There are approximately more than 50,000 terpenes that have been isolated from different plant species. Terpenes, composed of isoprene units (C_5H_8), play a variety of vital roles in plant species, including growth and development and defense against herbivores and environmental stress [8]. Terpenes possess a great variety of biological activities as antimicrobial, against cancer, malaria, and anti-inflammatory effects in acute and chronic inflammatory conditions like chronic obstructive pulmonary disease and osteoarthritis [9, 10].

Cyperus rotundus, a perennial plant, has several pharmacological activities, including antibacterial [11], antimutagenic [12], and anti-inflammatory [13]. Isocyperol, a sesquiterpene isolated from the rhizomes of *C. rotundus*, inhibited the production of NO and PGE2, decreased the levels pro-inflammatory interleukins (IL-1β and IL-6) and the monocyte chemotactic protein-1 (MCP-1), and suppressed the gene expression of iNOS and COX-2 in RAW-264 murine macrophages stimulated with lipopolysaccharide (LPS). In addition, isocyperol reduced the serum levels of NO, PGE2, and IL-6 in LPS-induced septic shock in mice, via suppression of the NF-κB and STAT3 signaling pathways [14]. *Dodonaea viscosa* induces gastroprotective [15], antibacterial [16], analgesic, and anti-inflammatory activities [17, 18]. Hawtriwaic acid, an ent-clerodane diterpene, was isolated from *D. viscosa* and showed anti-inflammatory activity on the murine ear edema induced with 12-O-tetradecanoylphorbol-13-acetate (TPA) by one or multiple applications. In both models, the compound diminished the edema [19]. Hawtriwaic acid at doses of 5, 10, and 20 mg/kg decreased knee inflammation in a murine model of monoarthritis induced with kaolin/carrageenan, by the reduction of serum levels of the pro-inflammatory interleukins Il-1β, IL-6, and TNF-α, and the increase in the serum levels of the anti-inflammatory interleukin IL-10 [20]. Ursolic acid, a pentacyclic triterpene found in many plant species, was identified for the first time in 1920 in the epicuticular waxes of apples. Ursolic acid exerts cytotoxic effects in various cancer cells by the inhibition of the STAT3 signaling pathway [21] and the induction of apoptosis [22]. Ursolic acid has protective effects on lung, kidney, liver, and brain, exerts anabolic effects on skeletal muscle [23], and induces antinociceptive activity in abdominal constriction test induced by acetic acid and the formalin test in mice [24]. Ursolic acid decreased the paw edema induced with carrageenan in rats [25], decreased the ear edema induced with Croton oil in mice [26], reduced the levels of iNOS, COX-2, IL-1β, IL-6, and TNF-α, and increased the level of IL-10 in macrophages stimulated with LPS [27].

The sesquiterpenes, vernomelitensin and onopordopicrin, isolated from *Onopordum illyricum* and the triterpene, Sootepin F, obtained from *Gardenia sootepensis*, decreased each NF-κB activity with IC_{50} values of 3.6, 8.6, and 20.3 μM, respectively [28, 29].

The pro-inflammatory enzymes: (1) inducible the nitric oxide synthase (iNOS), which is involved in the nitric oxide (NO) production, and the cyclooxygenase-2 (COX-2) involved in the prostaglandin production, are estimated in LPS-induced macrophages to evaluate the *in vitro* anti-inflammatory activity. IL-1 and TNF-α stimulate the production of NO. The inhibitory concentration 50 (IC_{50}) for these two pro-inflammatory enzymes has only been reported in some studies. The sesquiterpenes hydroxycostunolide (IC_{50} = 0.68 μM), costunolide (IC_{50} 0.3 μM), and artemorin (IC_{50} = 0.16 μM), obtained from *Inula montana*, showed similar or higher potency in the inhibition of NO, compared to that reported for the positive control dexamethasone (IC_{50} = 0.45–4.33 μM) [30]. Further studies are recommended to be performed with theses sesquiterpenes. Toxicological studies to guarantee their safety in long-term studies are also necessary. The *in vivo* studies evaluate swelling, redness, and pain mainly in rodents.

In the table is show the structure of 62 sesquiterpenes, 34 diterpenes, and 22 triterpenes with anti-inflammatory activity, isolated from 44 plant species, 1 sponge, and 2 corals.

Scientific name	Terpene	Structure	Anti-inflammatory assay	Results	Ref.
Sesquiterpenes					
Datura metel L.	**1)** 1β,5α,7β-Guaiane-4β,10α,11-triol **2)** 1α,5 α,7 α −11- Guaiene-2 α,3β,4α,10α,13-pentaol	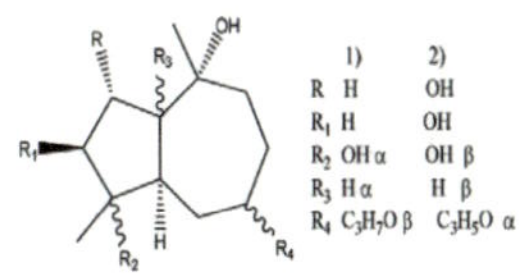	LPS-stimulated BV2 microglia cells.	The compounds **1** and **2**, tested at a concentration of 80 μM, each, inhibited the NO production by 37.5 and 46.0%, respectively.	[31]
Nardostachys chinensis Batal.	Nardosinanone N	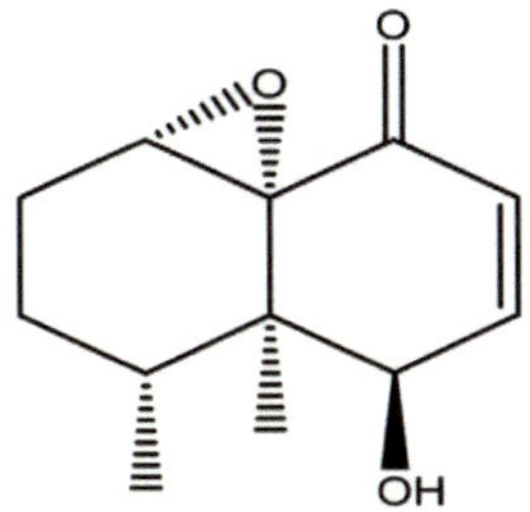	LPS-stimulated RAW 246.7 macrophages.	The compound tested at 30 μM inhibited the protein expression of iNOS and COX-2.	[32]
Chloranthus henryi Hemsl.	Shizukaol B	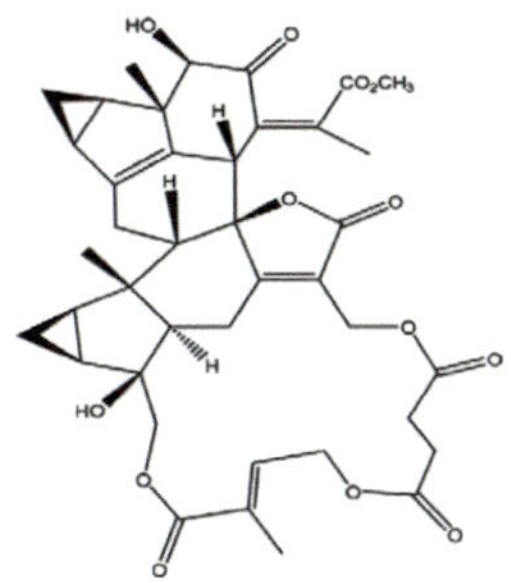	LPS-stimulated BV2 microglia cells.	The compound (12.5–50 μM) suppressed in concentration-dependent manner the protein expression of iNOS and COX_2, and decreased the production of NO, TNF-α, and IL-1β.	[33]

Scientific name	Terpene	Structure	Anti-inflammatory assay	Results	Ref.
Elephantopus mollis Kunth.	**1)** 8-*O*-Methacryloyl-elephanpane **2)** 2,4-bis-*O*-Methyl-8-*O*-methacryloyl-elephanpane **3)** 4-*O*-Ethyl-8-*O*-methacryloyl-elephanpane **4)** 2,5-Epoxy-2β-hydroxy-4α-methoxy-8α-(2-methyl-propenoyloxy)-10(14),11(13)-germacratrien-12,6α-olide	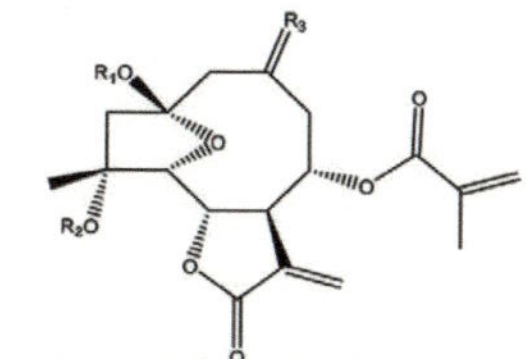 	LPS-stimulated RAW 264.7 macrophages.	The compounds inhibited the NO production with the following IC_{50} values (μM): **1** (2.09), **2** (2.18), **3** (4.06), **4** (4.82), **5** (14.34), **6** (59.97), **7** (0.57), **8** (2.17), **9** (2.02), **10** (1.95), **11** (11.25), **12** (1.09) **13** (1.21), **14** (6.95), and the positive control indomethacin (127.88).	[34]
	5) 2-*O*-Demethyl-tomenphantopin C **6)** Tomenphantopin C	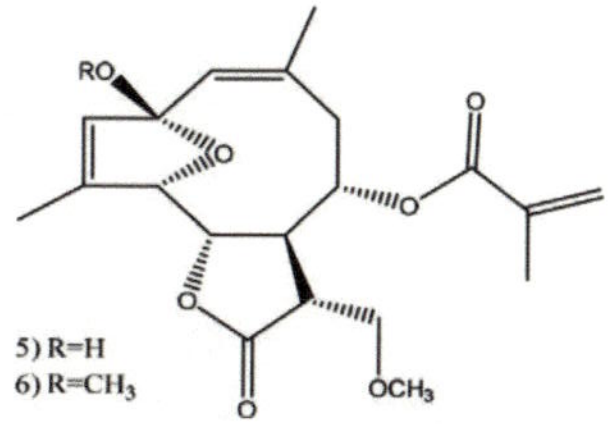 			
	7) Molephantin A	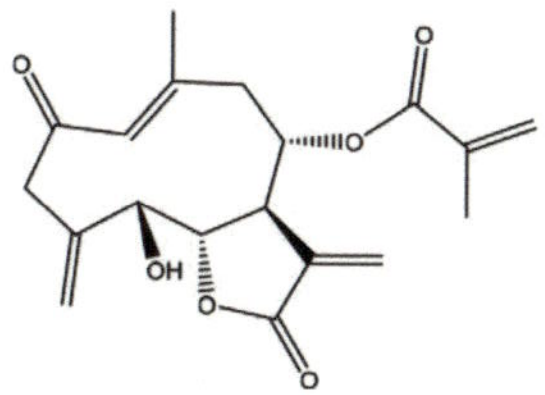 			
	8) Molephantin B	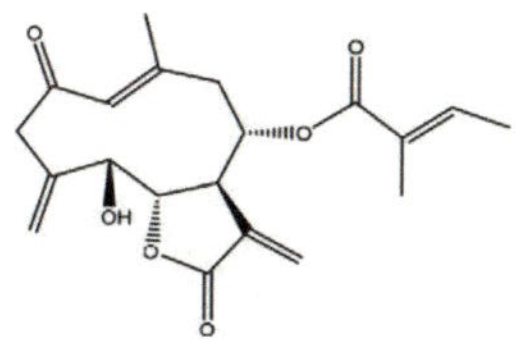 			

Scientific name	Terpene	Structure	Anti-inflammatory assay	Results	Ref.
	9) 2β-Methoxy-2-de-ethoxy-phantomolin **10)** 2β-Methoxy-2-deethoxy-8-*O*-deacyl-phantomolin-8-*O*-tiglinate	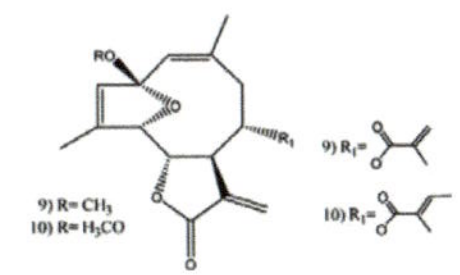			
	11) Scabertopin **12)** Isodeoxyelephantopi	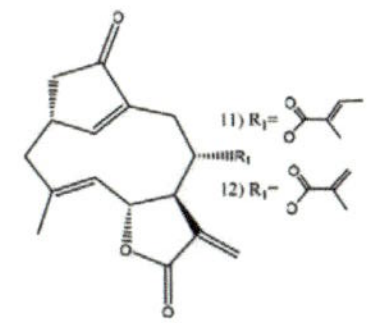			
	13) Molephantin	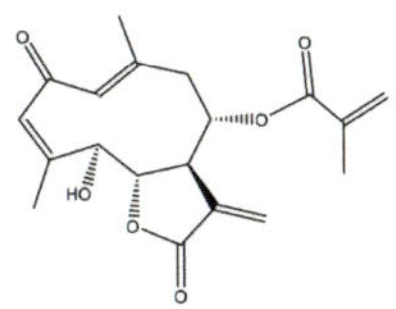			
	14) 2, 5-Epoxy- 2β-hydroxy-4α-methoxy-8α-[(E)-2-methylbut-2-enoyloxy] -4(15), 10(14), 11(13)-germacratrien-12,6α-olide	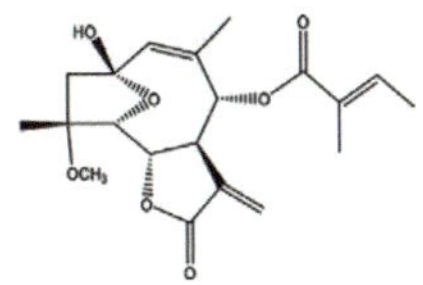			
Chloranthus holostegius (Hand-Mazz) S. J. Pei & Shan.	**1)** Chololactone E **2)** Chololactone F	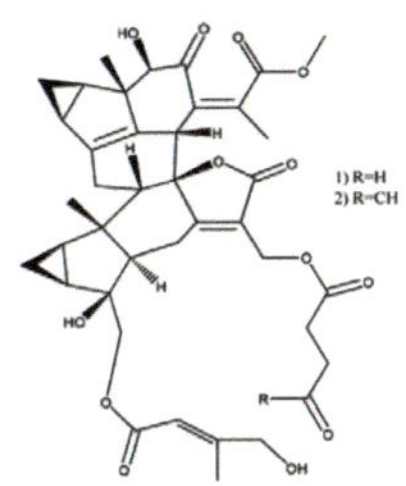	LPS-stimulated RAW 264.7 macrophages.	All compounds reduced NO production with IC_{50} (μM) values: **1** (8.5), **2** (6.9), **3** (3.5), **4** (4.4), and positive control dexamethasone (0.45).	[35]

http://dx.doi.org/10.5772/intechopen.73215

Scientific name	Terpene	Structure	Anti-inflammatory assay	Results	Ref.
	3) Chololactone G	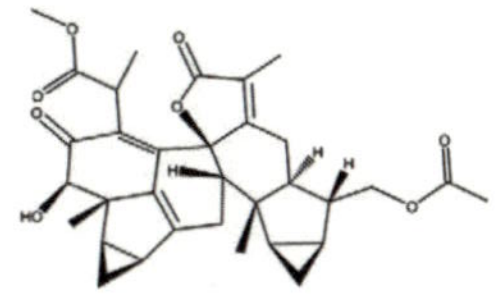			
	4) Chololactone H	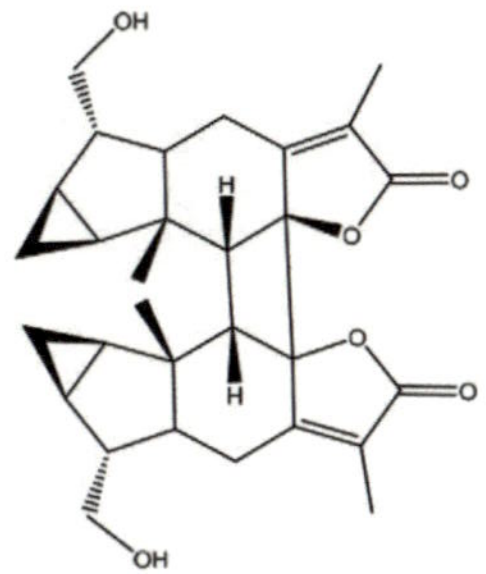			
Smenospongia cerebriformis Duchassaing & Michelotti.	**1)** Smenohaimien A **2)** Smenohaimien B	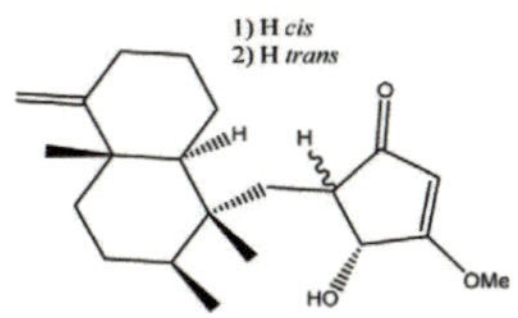	LPS-stimulated BV2 microglial cells.	The compounds inhibited NO production **1** (IC_{50} 30.13 μM), **2** (IC_{50} 28.33 μM), **3** (IC_{50} 24.37 μM), **4** (IC_{50} 24.44 μM), and positive control l-NMMA (22.1 μM).	[36]
	3) Smenohaimien E	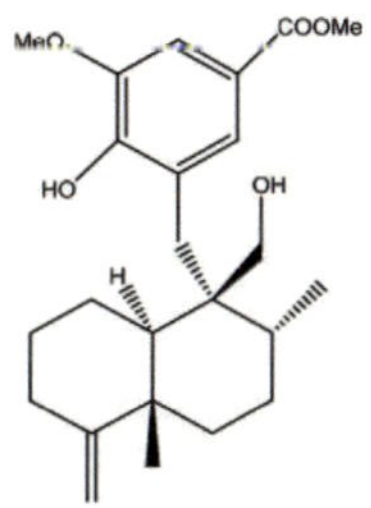			

Scientific name	Terpene	Structure	Anti-inflammatory assay	Results	Ref.
	4) Ilimaquinone	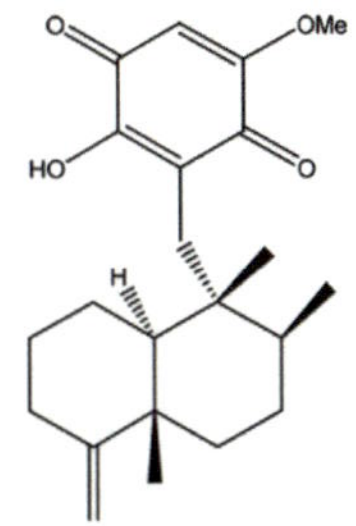			
Inula montana L.	**1)** Artemorin	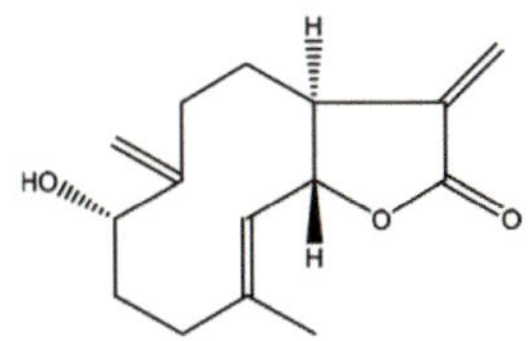	LPS-stimulated RAW 264.7 macrophages.	The compounds inhibited the NO production with the following IC_{50} values (μM): **1** (0.16), **2** (0.3), **3** (0.68), and the positive control dexamethasone (4.33).	[30]
	2) Costunolide **3)** 9β-Hydroxycostunolide	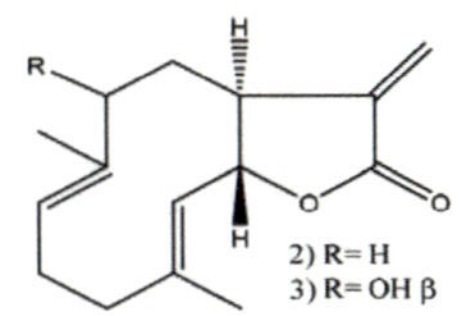			
Ainsliaea yunnanensis Franch.	Zaluzanin C	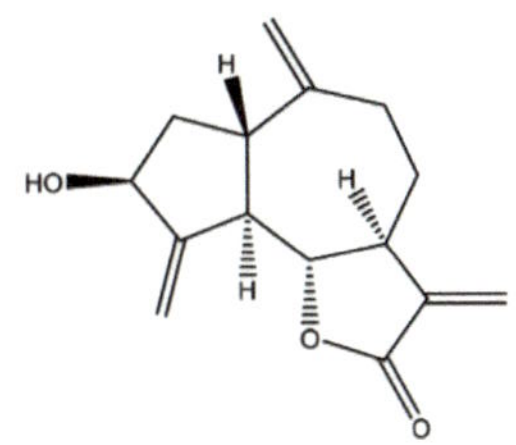	LPS-stimulated RAW 264.7 macrophages.	The compound (IC_{50} 6.54 μM) reduced the NO production with similar potency, compared to the positive control amino guanidine (IC_{50} 7.30 μM).	[37]

Scientific name	Terpene	Structure	Anti-inflammatory assay	Results	Ref.
Chloranthus japonicus Siebold.	Chlorajaponol B	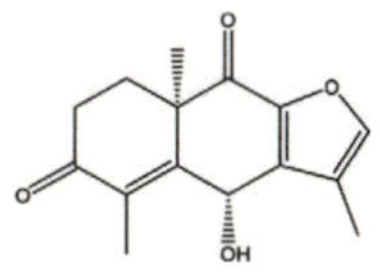	LPS-stimulated RAW 264.7 macrophages.	The compound decreased the NO production (IC_{50} 9.56 μM).	[38]
Neurolaena lobata (L.) Cass.	**1)** Neurolenin B **2)** Neurolenin C **3)** Neurolenin D	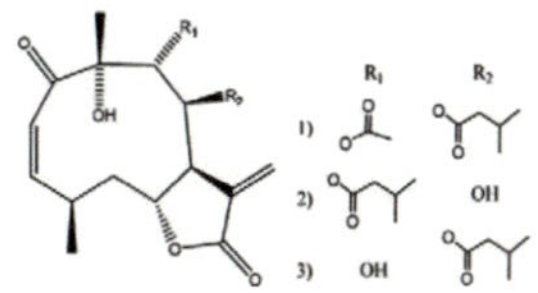	LPS-stimulated monocytes.	The compounds **1–5** are potent inhibitors of TNF-α production **1** (IC_{50} 2.32 μM), **2** and **3** (IC_{50} 1.10 μM), **4** (IC_{50} 0.17 μM), and **5** (IC_{50} 1.30 μM).	[39]
	4) Lobatin B	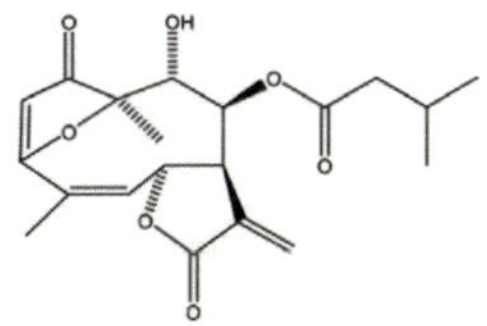			
	5) 9α-Hydroxy-8β-isovalerianyloxycalyculatolide	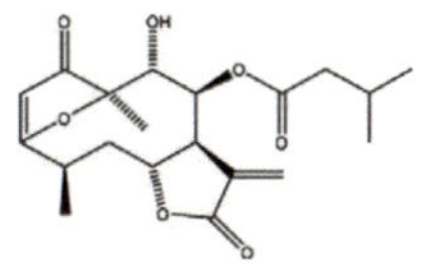			
Onopordum illyricum L.	Vernomelitensin	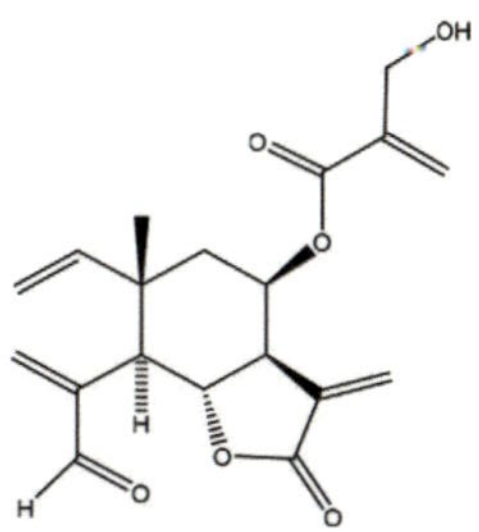	Anti-NF-κB activity was evaluated in the NIH-3 T3-KBF-Luc cell line. The antiSTAT3 activity was analyzed in HeLa-STAT3-luc cell line. The activation of the Nrf2 pathway was analyzed in the HaCaT-ARE-Luc cell line. The protein concentration in the	Vernomelitensin decreased the activity of NF-κB (IC_{50} 3.6 μM), STAT3 (IC_{50} 27.9 μM), and Nrf2 (IC_{50} 1.1 μM) Onopordopicrin decreased the activity of NF-κB (IC_{50} 8.6 μM), STAT3 (IC_{50} 15.3 μM), and Nrf2 (IC_{50} 2.2 μM).	[28]

Scientific name	Terpene	Structure	Anti-inflammatory assay	Results	Ref.
	Onopordopicrin	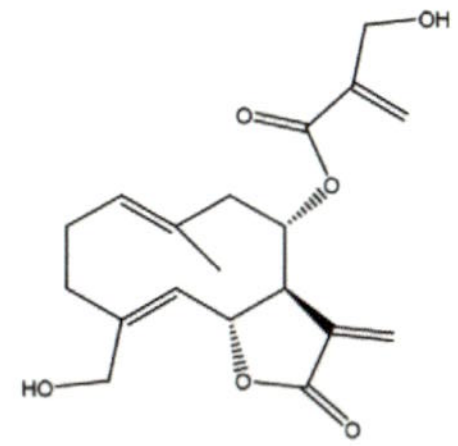	cell extracts were measured by the Bradford method (BioRad).		
Lychnophora passerina (Mart ex DC) Gardner.	Goyazensolide	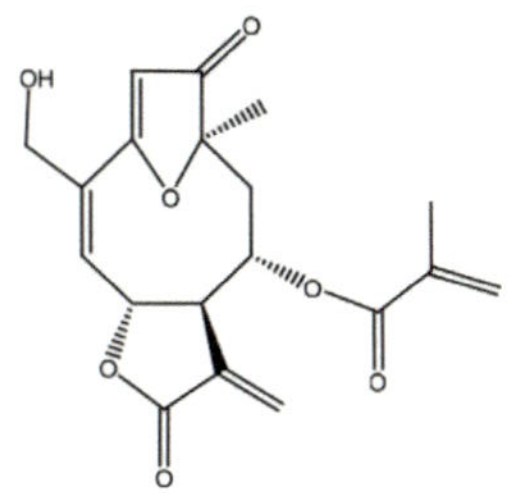	Carrageenan-induced paw edema assay.	Goyazensolide (10 mg/kg orally) reduced carrageenan-induced paw edema assay in the second hour from the assay. This sesquiterpene lactone presented similar effect to indomethacin.	[40]
Lindera strychnifolia (Sieb. & Zucc.) Fern Vill.	**1)** Linderolide O **2)** Linderolide P **3)** Linderolide Q **4)** Linderolide R	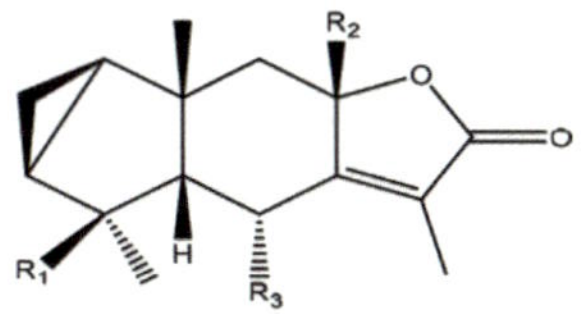 1) R_1=OCH_3; R_2=H 2) R_1=OH; R_2=OH; R_3=H 3) R_1=OH; R_2=OH;R_3=OH 4) R_1=OH; R_2=OH;R_3=OAc	LPS-stimulated RAW 264.7 macrophages.	The compounds inhibited the NO production with the following IC_{50} values (μM): **1** (6.3), **2** (9.6), **3** (14.6), **4** (12.8), **5** (15.4), **6** (9), and the positive control amino guanidine (18.4).	[41]
	5) Menelloide C	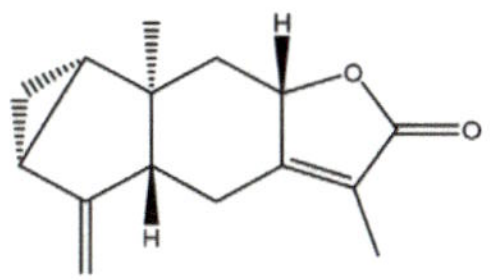			

Scientific name	Terpene	Structure	Anti-inflammatory assay	Results	Ref.
	6) Parvigemone	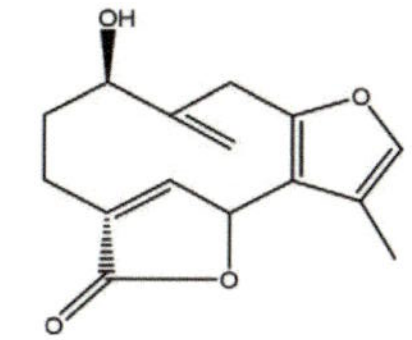			
Artemisia austroyunnanensis Y. Ling & Y. R. Ling.	**1)** Austroyunnane B	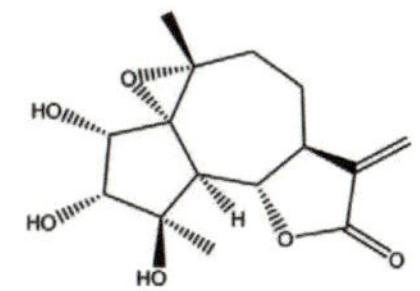	LPS-stimulated RAW 264.7 macrophages.	The compounds inhibited the NO production with the following IC_{50} values (μM): **1** (4.20), **2** (10.67), **3** (5.10), **4** (8.77), **5** (4.21), **6** (2.57), and the positive control L-NMMA (24.95).	[42]
	2) Austroyunnane C	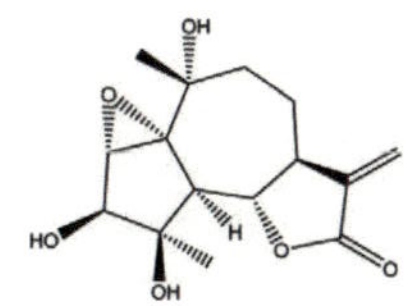			
	3) Austroyunnane D	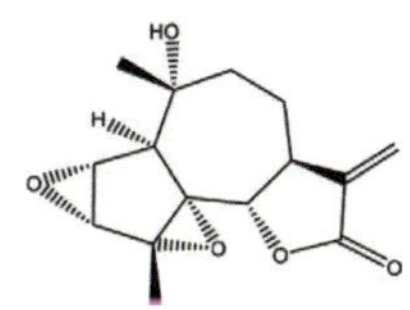			
	4) Artecaninhydrate	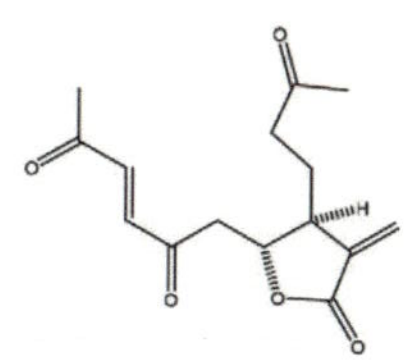			

Scientific name	Terpene	Structure	Anti-inflammatory assay	Results	Ref.
	5) 3α- Chloro-4β,10α-dihydroxy-1β,2β-epoxy-5α,7α*H*-guai-11(13)-en-12,6α-olide	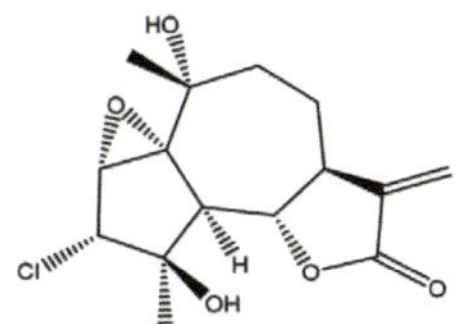			
	6) 15,16-Reynosin	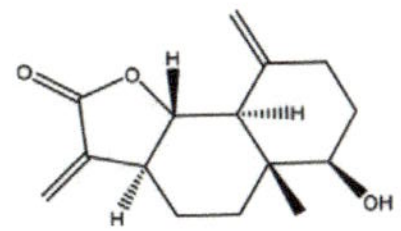			
Curcuma wenyujin Y. H. Chen & C. Ling.	**1)** 7α,11-Epoxy-6α-methoxy-carabrane-4,8-dione **2)** 8,11-Epidioxy-8-hydroxy-4-oxo-6-carabren **3)** Isogermafurenolid **4)** Hydroxyisogermafurenolide **5)** Curcumenone **6)** Neocurdione **7)** Curdionolide B **8)** Isocurcumol **9)** 4-Epicurcumol	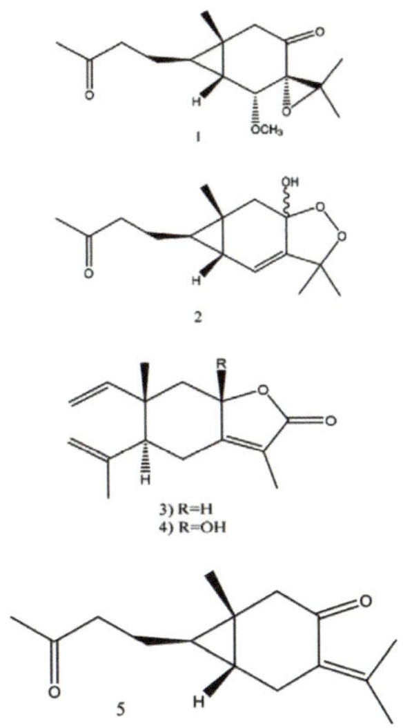 	LPS-stimulated RAW 264.7 macrophages.	The compounds inhibited the NO production with the following IC_{50} values (μM): **1** (47.05), **2** (25.36), **3** (30.62), **4** (59.97), **5** (34.69), **6** (24.18), **7** (14.50), **8** (22.36), **9** (17.26), and the positive control hydrocortisone (64.34).	[43]

Scientific name	Terpene	Structure	Anti-inflammatory assay	Results	Ref.
		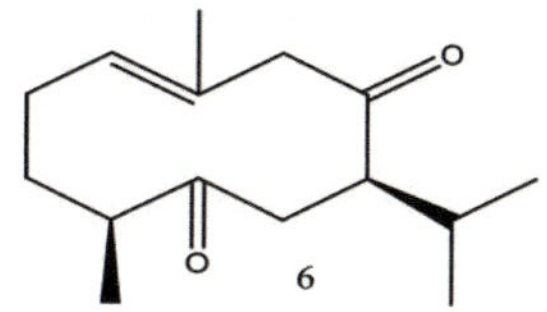 6			
		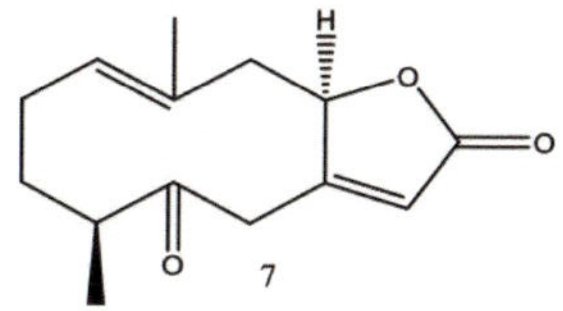 7			
		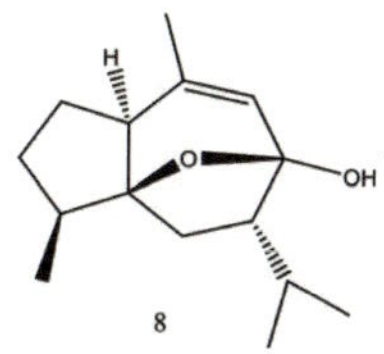 8 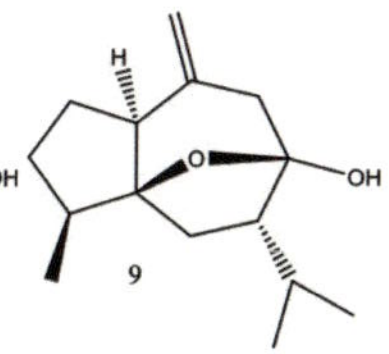 9			
Pulicaria undulata (L.) C.A. Mey.	1β,4β-Dihydroxy-5α*H*,7α*H*,8α-guaia-10(14),11(13)-dien-8α,12-olide	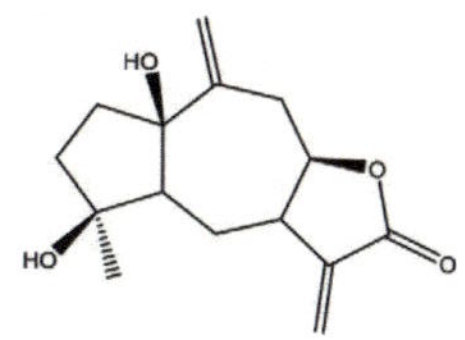	LPS-stimulated RAW 264.7 macrophages.	The compound inhibited the NO production with IC_{50} 7.2 μM.	[44]
Diterpenes					
Thuja orientalis L.	15-nor-14-Oxolabda-8(17),13(16)-dien-19-oic acid	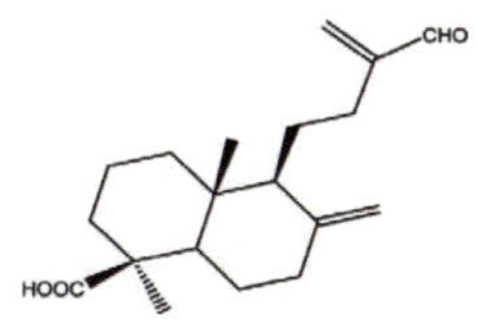	LPS-stimulated RAW 264.7 macrophages.	The compound reduced the expression of iNOS and COX-2 (IC_{50} 1: 3.56 μM). Curcumin was used as a positive control.	[45]

Scientific name	Terpene	Structure	Anti-inflammatory assay	Results	Ref.
Pseudopterogorgia acerosa Pallas.	Pseudopterane	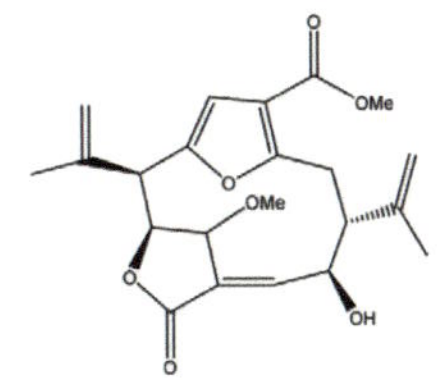	LPS-stimulated peritoneal macrophages.	This compound (25 μM) inhibited the expression and secretion of TNF-α, IL-6, IL-1β, NO, IP-10, COX-2 iNOS and monocyte chemoattractant protein-1 (IC_{50} = 12.25).	[46]
Aconitum laciniatum (Brühl) Stapf.	14-*O*-Acetylneoline	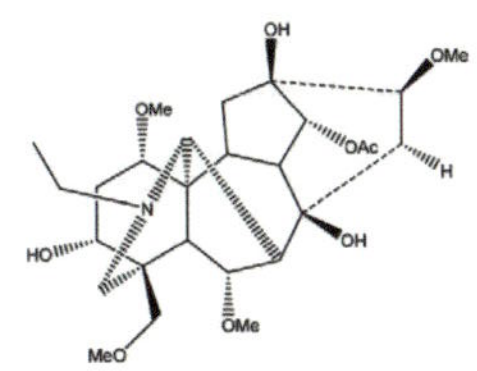	TNBS-induced colitis model in mice.	At doses 10, 20 and 50 μg/ mouse showed significant protection against different parameters of colitis inflammation.	[47]
Aconitum koreanum Rapaics.	Acanthoic acid	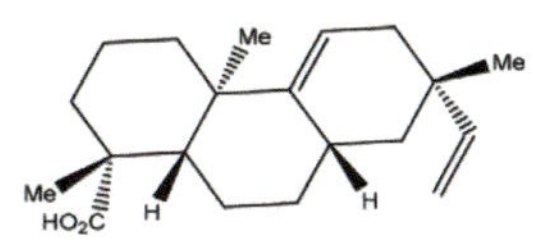	LPS-stimulated RAW 264.7 peritoneal macrophages. Ear edema in mice induced by TPA.	The compound (10 μM) decreased the levels of IL-1β, IL-18, TNF-α, and IFN-γ. Decreased ear edema (0.5 μg/ ear), and indomethacin was used as the positive control (0.5 μg/ear).	[48]
Hedychium coronarium J. Koening.	7β-Hydroxycalcaratarin A	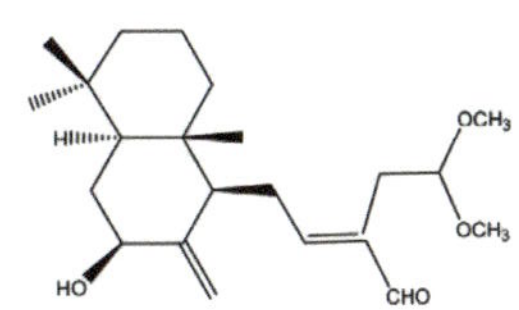	Measurement of superoxide anion generation.	Inhibition (IC_{50} 4.52 μg/mL) of superoxide anion generation by human neutrophils.	[49]
Physalis angulata L.	**1)** Physangulatoside A **2)** Physangulatoside F	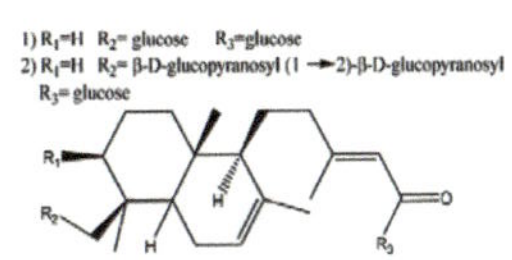	LPS-stimulated RAW 264.7 macrophages.	Compounds displayed inhibitory effects against NO production with IC_{50} values of 15.9 μM and 60.7 μM, respectively.	[50]

Scientific name	Terpene	Structure	Anti-inflammatory assay	Results	Ref.
Croton stellatopilosus H. Ohba.	**1)** Plaunotol **2)** Plaunolide **3)** Plaunol E	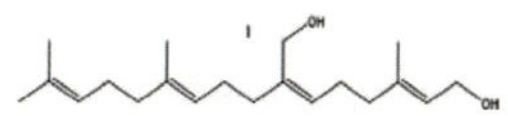 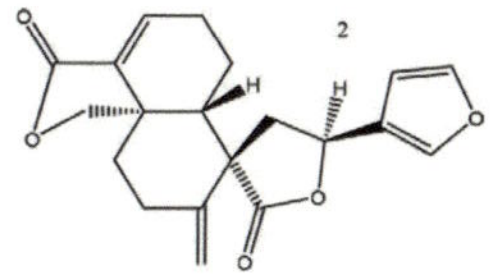 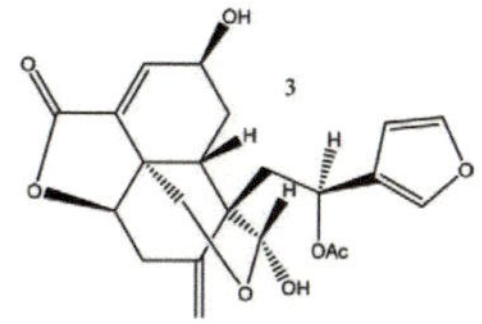	LPS-stimulated RAW 264.7 macrophages.	Plaunotol, plaunolide and plaunol E showed IC_{50} values of 3.41, 17.09, and 2.79 μM, respectively. Plaunotol suppressed expression of the iNOS gene. Plaunolide decreased the expression of the COX-1, COX-2 and iNOS genes. Plaunol E inhibited the expression of the COX-2.	[51]
Dodonaea viscosa Jacq.	Hawtriwaic acid	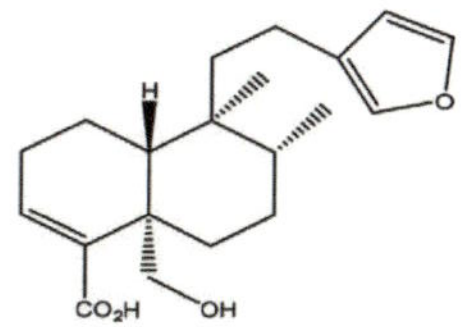	Kaolin/carrageenan-induced monoarthritis model. Ear edema induced with TPA.	After 10 days of treatment with different doses of hawtriwaic acid (5, 10, 20 mg/kg) decreased knee inflammation in a range of 40–70%. Hawtriwaic acid decreased the levels of proinflammatory cytokines IL-1β, IL-6, and TNF-α.	[20]
Callicarpa randaiensis Hayata.	Randainins D	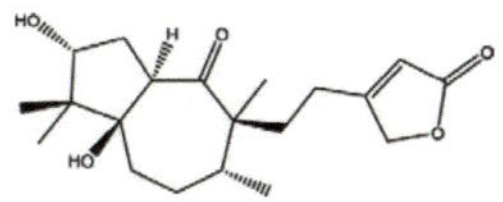	Measurement of superoxide anion generation.	The compound exhibited moderate inhibition of superoxide anion generation with an IC_{50} value of 21.5 μM.	[52]
Siegesbeckia pubescens (Makino) Makino.	**1)** Methyl *ent*-16α*H*-17-Hydroxy-kauran-19-oate **2)** Kirenol	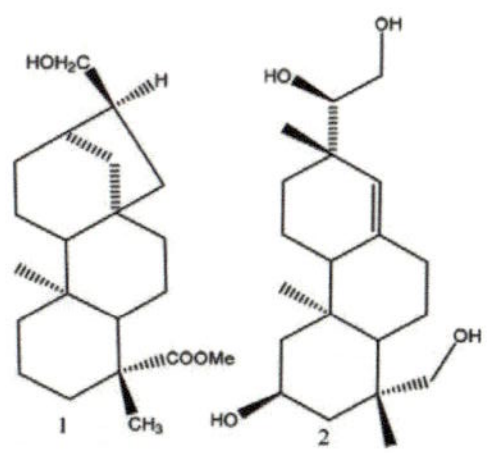	LPS-stimulated BV2 microglia cells.	Each compound, tested at 100 μM, decreased the expression of iNOS and COX-2 by 20.5% and 22.5%, respectively.	[53]

Scientific name	Terpene	Structure	Anti-inflammatory assay	Results	Ref.
Leonurus japonicus Houtt.	Hispanone	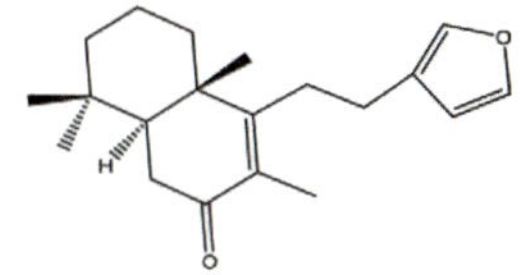	Inhibition of superoxide anion generation and elastase release.	The compound decreased the superoxide anion with IC_{50} = 8.48 μM.	[54]
Cephalotaxus lanceolata K.M. Feng.	**1)** Lanceolatins C **2)** Gongshanolide	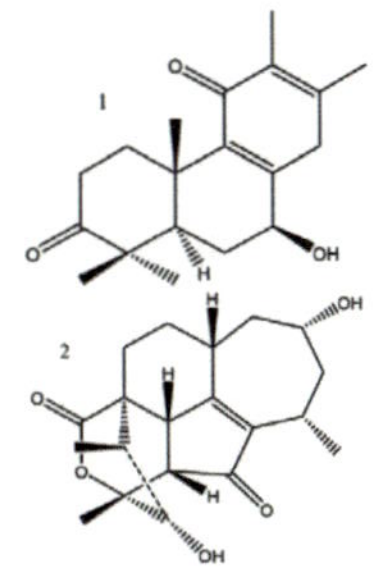	LPS-stimulated RAW 264.7 macrophages.	The compounds decreased the NO production with IC_{50} values of 10.79 μM and 12.73 μM, respectively. .	[55]
Ocimum labiatum (N.E. Br.) A.J. Paton.	Labda-8(17),12E,14-triene-2R,18-diol	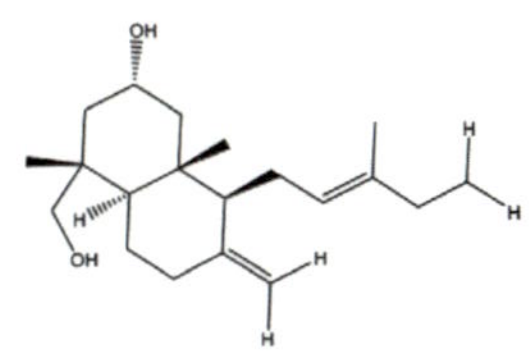	Cytometric bead array (CBA) technique. The effect on phytohemagglutinin (PHA)-induced nitric oxide (NO) production in peripheral blood mononuclear cells (PBMCs) was also assessed.	The compound tested at 25 μg/ml inhibited the production of pro-inflammatory cytokines IL-2, IL-4, IL-6, and IL-17A. The compound decreased the NO production with IC_{50} value of 70 μM.	[56]
Cunninghamia konishii Hayata.	**1)** Konishone **2)** Hinokiol **3)** 12-Hydroxy-6,7-secoabieta-8,11,13-triene-6,7-dial	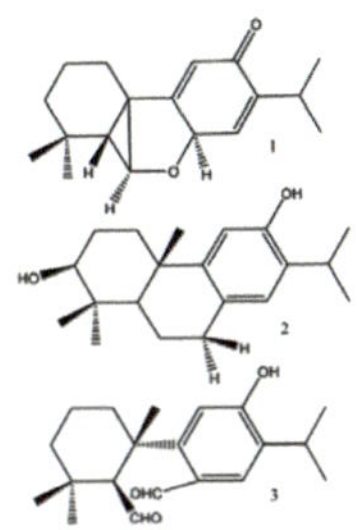	LPS-stimulated RAW 264.7 macrophages.	The compounds inhibited the NO production with the following IC_{50} values (μM): **1** (9.8), **2** (7.9), and **3** (9.3).	[57]

Scientific name	Terpene	Structure	Anti-inflammatory assay	Results	Ref.
Caesalpinia pulcherrima (L.) Sw.	**1)** Pulcherrimin B **2)** Pulcherrimin D	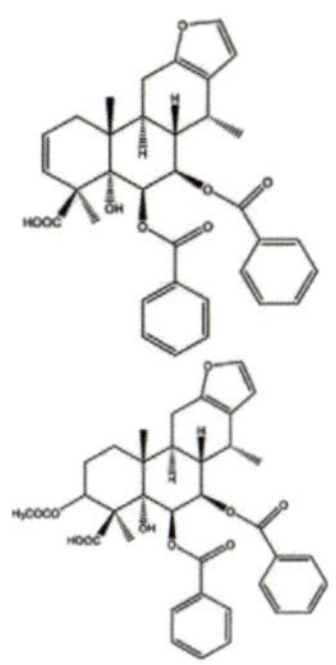	Evaluated via the oxidative burst assay using a luminol-amplified chemiluminescence technique.	The compounds inhibited the production of reactive oxygen species with IC_{50} values of 15.30 μM and 8 μM, respectively.	[58]
Lobophytum crassum von Merenzeller.	**1)** Lobocrasol A **2)** Lobocrasol B	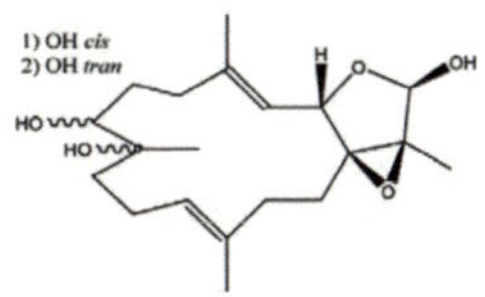	Were evaluated using NF-kB luciferase and reverse transcription polymerase chain reaction (RT-PCR).	Significantly inhibited TNF-α induced NF-kB transcriptional activity in HepG2 cells in a dose-dependent, with IC_{50} values of 6.30 and 6.63 μM.	[59]
Euphorbia helioscopia L.	Jatrophane	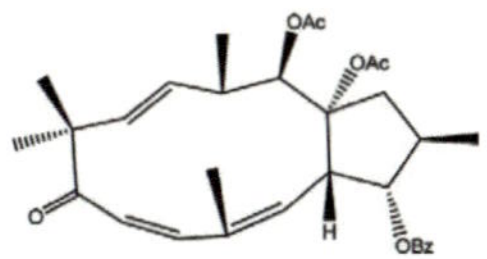	LPS-stimulated BV-2 microglial cells.	The compound inhibited the NO production (IC_{50} = 73.7 μM).	[60]
Erythrophleum ivorense A. Chev.	Erythroivorensin	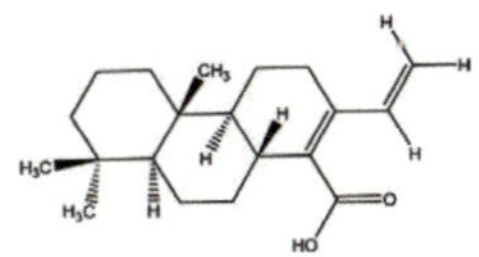	The carrageenan paw edema model in chicken.	The compound (300 mg/kg orally) showed anti-inflammatory effect with similar activity compared to diclofenac (100 mg/Kg).	[61]

Scientific name	Terpene	Structure	Anti-inflammatory assay	Results	Ref.
Crinipellis species.	**1)** Crinipellin E **2)** Crinipellin F **3)** Crinipellin G **4)** Crinipellin H	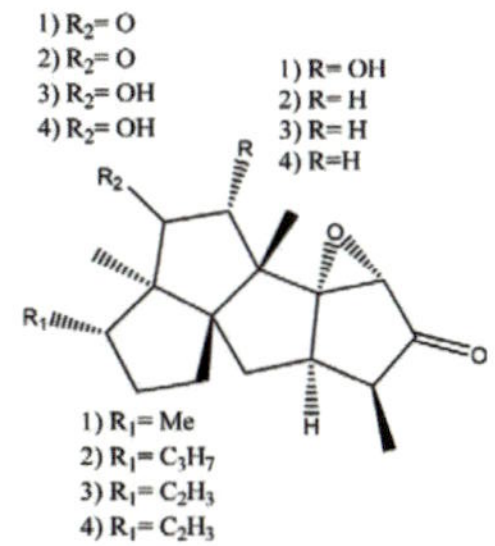	LPS/IFN-γ induced CXCL10 promoter activity in transiently transfected human MonoMac6 cells.	Reduce the inducible expression of pro-inflammatory genes in LPS/IFN-γ stimulated human MonoMac6 cells (IC_{50} values of 15, 1.5, 1.3, and 3.15 μM, respectively).	[62]
Salvia divinorum Epling & Játiva.	Salvinorin A	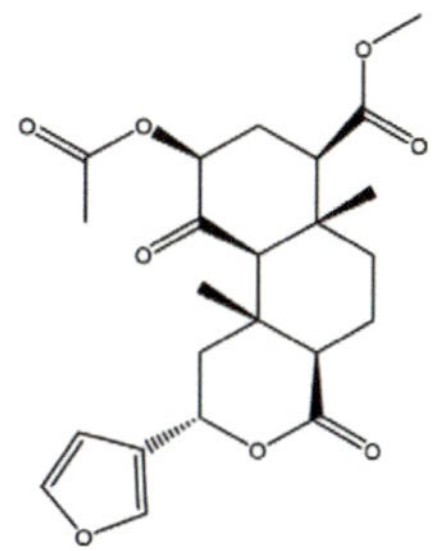	Calcium ionophore-stimulated peritoneal macrophages, and in carrageenan-induced pleurisy in rats.	Inhibited the production of leukotrienes. In carrageenan-induced pleurisy in rats, Salvinorin A significantly inhibited LTB_4 production in the inflammatory exudates, along with reducing the phlogistic process in the lung.	[63]
Casearia sylvestris Sw.	**1)** Caseargrewiin F **2)** Casearin B	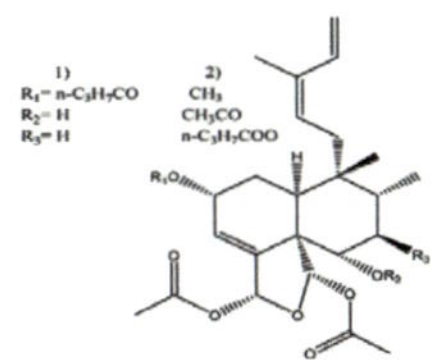	Carrageenan-induced paw edema.	At doses 10 and 100 mg/kg showed a significant reduction of the carrageenan-induced paw edema in rats compared to indomethacin.	[64]

Scientific name	Terpene	Structure	Anti-inflammatory assay	Results	Ref.
Triterpenes					
Anopyxis klaineana Pierre (Engl.).	**1)** 3,23-Dioxotirucalla-7,24-dien-21-oic acid **2)** Methyl angolensate	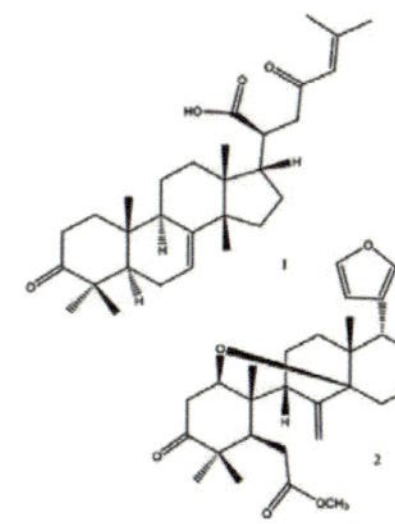	The PGE_2 assay method, sequential competitive enzyme immunoassay with RAW 264.7.	The IC_{50} values were 3.63 μM and 10.23 μM for compounds **1** and **2**, respectively. Cortisone was used as reference standard.	[65]
Bursera copallifera (DC.) Bullock.	**1)** 3-Epilupeol formiate **2)** α-Amyrin acetate **3)** 3-Epilupeol acetate **4)** Lupenone **5)** 3-Epilupeol **6)** α-Amyrin	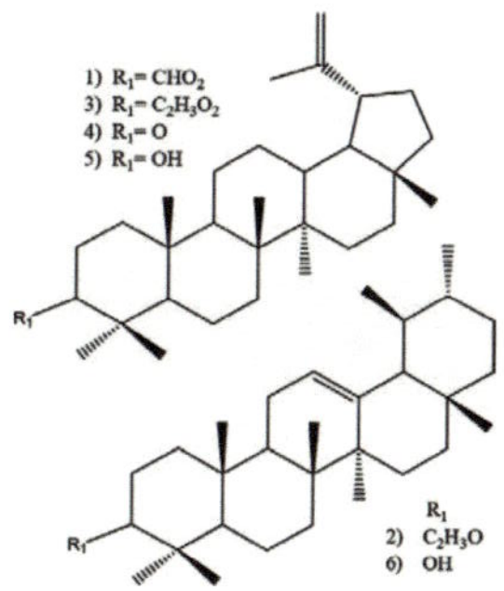	TPA-induced ear edema in mice, and LPS-stimulated RAW 264.7 macrophages.	The anti-inflammatory effect (1 mg/ear) on ear edema induced by TPA were: **1)** 55.14%, **2)** 62.12%, **3)** 49.35%, **4)** 57.25%, **5)** 66.39, and **6)** 25%. The compounds inhibited the NO production with IC_{50} values (μM): **1** (43.31), **2** (22.57), **3** (31.13), **4** (20.8), **5** (15.5), and **6** (8.98).	[66]
Euonymus carnosus Hemsl.	**1)** 3α-Acetoxy-28-hydroxy-30-norlupane-18(19)-en-20-one **2)** 3α-Hydroxylup-20(29)-en-30-oic acid	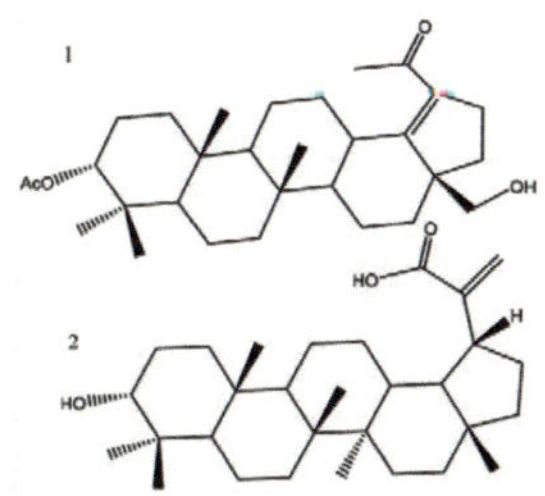	LPS-stimulated BV-2 microglial cells.	The compounds inhibited the NO production with IC_{50} values (μM): **1** (5.99) and **2** (8.47).	[67]

Scientific name	Terpene	Structure	Anti-inflammatory assay	Results	Ref.
Chaenomeles sinensis (Thouin) Koehne.	**1)** Sinenic A acid **2)** 3β-*O*-cis-Feruloyl-2α,19α-dihydroxy-urs-12-en-28-oic acid **3)** 3β-*O*-cis-caffeoylbetulin **4)** Betulinic acid	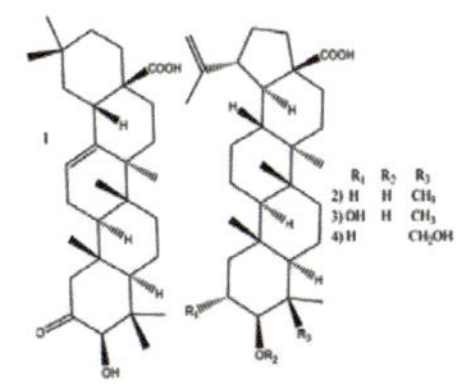	LPS-stimulated microglia BV2 cells.	The compounds inhibited the NO production with the following IC_{50} values (μM): **1** (17.8), **2** (4.5), **3** (14.5), and **4** (13.4).	[68]
Gardenia sootepensis Hutch.	**1)** Sootepin F **2)** Sootepin G **3)** Sootepin H	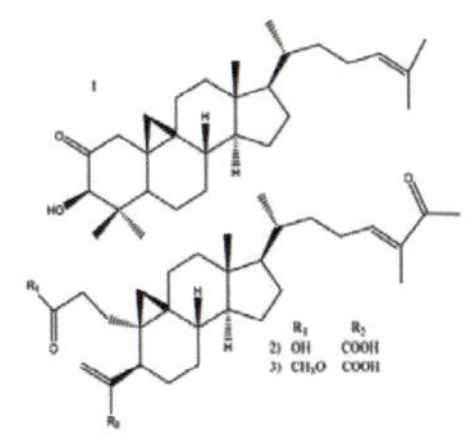	LPS-stimulated macrophages RAW 264.7.	The compounds inhibited the NO production with the following IC_{50} values (μM): **1** (43.6), **2** (18.4), and **3** (15). The compounds inhibited the NF-kB activity with the following IC_{50} values (μM): **1** (20.3), **2** (>100), and **3** (42.3).	[29]
Protium paniculatum Engl.	**1)** α-Amyrin **2)** α-Amyrone **3)** Maniladiol	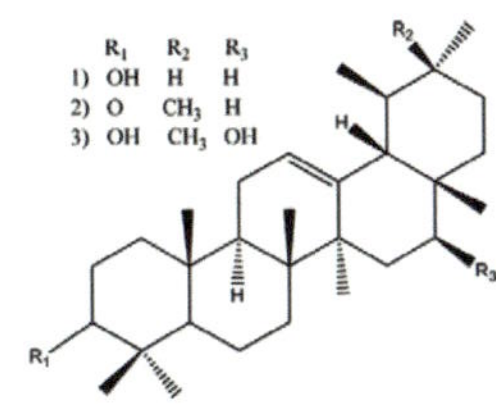	LPS-stimulated J774A.1 macrophages.	The compounds tested at 1 μg/ml inhibited the NO production by 98.34% **1**), 99.86%, **2**), and 96.05% **3**).	[69]
Quercus serrata var. brevipetiolata (A. DC.) Nakai.	**1)** 3,23-*O*-Methyl butyrate-2,3,19,23-tetrahydroxy-urs-12-en-28-oic acid β-D- glucopyranosyl ester **2)** 3,23-*O*-Methyl butyrate–2,3,19,23-tetrahydroxy-olean-12-en-28-oic acid β-D-glucopyranosyl ester **3)** 23-Acetoxy-2,3,19 -trihydroxyurs-12-en-28-oic acid	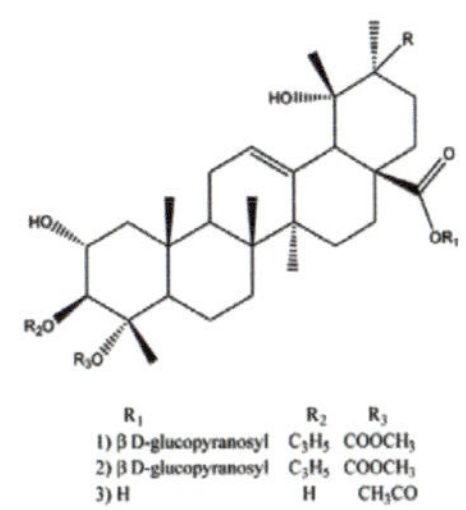	LPS-induced NO production and interleukins pro-inflammatory.	The compounds inhibited the NO production with the following IC_{50} values (μM): **1** (8.2) μM, **2** (12.8), and **3** (19.1).	[70]

3. Conclusion

This book chapter indicates that there has been an increase in the search of terpenes with anti-inflammatory activity in recent years. This fact indicates that terpenes are a topic of interest. The possible mechanisms involved in the anti-inflammatory effects of the terpenes are pointing out on the inhibition of NF-κB, TNF-α, PGE2, and pro-inflammatory cytokines such as IL-6. NF-κB is one of the current targets for the development of new anti-inflammatory drugs [71]. In addition, the molecular targets of terpenes are highly desirable to find target-specific anti-inflammatory drugs. The mechanism of action of many terpenes remains to be studied. The combination of terpenes with high anti-inflammatory activity and with studied mechanism of action and currently used drugs could be another strategy for further anti-inflammatory therapy.

Author details

Roberto José Serrano Vega[1], Nimsi Campos Xolalpa[1], Angel Josabad Alonso Castro[2], Cuauhtémoc Pérez González[1], Julia Pérez Ramos[1] and Salud Pérez Gutiérrez[1]*

*Address all correspondence to: msperez@correo.xoc.uam.mx

1 Universidad Autónoma Metropolitana-Xochimilco, Delegación Coyoacan, Ciudad de México, México

2 Departamento de Farmacia, División de Ciencias Naturales y Exactas, Universidad de Guanajuato, Guanajuato, GTO, México

References

[1] Shafi O. Inverse relationship between Alzheimer's disease and cancer, and other factors contributing to Alzheimer's disease: A systematic review. BMC Neurology. 2016;**16**(1): 236. DOI: 10.1186/s12883-016-0765-2

[2] Ben-Neriah Y, Karin M. Inflammation meets cancer, with NF-[kappa] B as the matchmaker. Nature Immunology. 2011;**12**:715-723. DOI: 10.1038/ni.2060

[3] Markiewski MM, Lambris JD. The role of complement in inflammatory diseases from behind the scenes into the spotlight. The American Journal of Pathology. 2007;**171**(3):227-240. DOI: 10.2353/ajpath.2007.070166

[4] Yang HZ, Wang JP, Mi S, Liu HZ, Cui B, Yan HM, Lu W. TLR4 activity is required in the resolution of pulmonary inflammation and fibrosis after acute and chronic lung injury. The American Journal of Pathology. 2012;**180**:275-292. DOI: 10.1016/j.ajpath.2011.09.019

[5] Gordon S. Alternative activation of macrophages. Nature Reviews Immunology. 2016; **3**(1):23-36. DOI: 10.1038/nri978

[6] Pereira-Leite C, Nunes C, Jamal SK, Cuccovia IM, Reis S. Nonsteroidal anti-inflammatory therapy: A journey toward safety. Medicinal Research Reviews. 2016;**37**:802-859. DOI: 10.1002/med.21424

[7] Javadi B, Sahebkar A. Natural products with anti-inflammatory and immunomodulatory activities against autoimmune myocarditis. Pharmacological Research. 2017;**124**:34-42. DOI: 10.1016/j.phrs.2017.07.022

[8] Mewalal R, Rai DK, Kainer D, Chen F, Külheim C, Peter GF, Tuskan GA. Plant-derived terpenes: A feedstock for specialty biofuels. Trends in Biotechnology. 2017;**35**(3):227-240. DOI: 10.1016/j.tibtech. 2016-08.003

[9] Rufino A, Ribeiro M, Judas F, Salgueiro L, Lopes M, Cavaleiro C, Mendes A. Anti-inflammatory and chondroprotective activity of (+)-α-pinene: Structural and enantiomeric selectivity. Journal of Natural Products. 2014;**77**(2):264-269. DOI: 10.1021/np400828x

[10] Ma J, Xu H, Wu J, Qu C, Sun F, Xu S. Linalool inhibits cigarette smoke-induced lung inflammation by inhibiting NF-κB activation. International Immunopharmacology. 2015; **29**(2):708-713. DOI: 10.1016/j.intimp.2015.09.005

[11] Kilani S, Sghaier M, Limem I, Bouhlel I, Boubaker J, Bhouri W, Skandrani I, Neffatti A, Ben-Ammar R, Dijoux-Franca M, Ghedira K, Chekir-Ghedira L. In vitro evaluation of antibacterial, antioxidant, cytotoxic and apoptotic activities of the tubers infusion and extracts of *Cyperus rotundus*. Bioresource Technology. 2008;**99**(18):9004-9008. DOI: 10.1016/j.biortech.2008.04.066

[12] Kilani S, Ammar R, Bouhlel I, Abdelwahed A, Hayder N, Ben-Chibani J, Mahmoud A, Ben-Chibani L, Chekir-Ghedira K. Investigation of extracts from (Tunisian) *Cyperus rotundus* as antimutagens and radical scavengers. Environmental Toxicology and Pharmacology. 2005;**20**(3):478-484. DOI: 10.1016/j.etap.2005.05.012

[13] Dang G, Parekar R, Kamat S, Scindia A, Rege N. Antiinflammatory activity of *Phyllanthus emblica, Plumbago zeylanica* and *Cyperus rotundus* in acute models of inflammation. Phytotherapy Research;**25**(6):904-908. DOI: 10.1016/S0014-5793(01)03293-8

[14] Seo Y, Jeong M, Lee K, Jang DS, Choi J. Isocyperol, isolated from the rhizomes of *Cyperus rotundus*, inhibits LPS-induced inflammatory responses via suppression of the NF-κB and STAT3 pathways and ROS stress in LPS-stimulated RAW 264.7 cells. International Immunopharmacology. 2016;**38**:61-69. DOI: 10.1016/j.intimp.2016.05.017

[15] Arun M, Asha V. Gastroprotective effect of *Dodonaea viscosa* on various experimental ulcer models. Journal of Ethnopharmacology. 2008;**118**(3):460-465. DOI: 10.1016/j.jep.2008.05.026

[16] Khurram M, Khan M, Hameed A, Abbas N, Qayum A, Inayat H. Antibacterial activities of *Dodonaea viscosa* using contact bioautography technique. Molecules. 2009;**14**(3):1332-1341. DOI: 10.3390/molecules14031332

[17] Khalil N, Sperotto J, Manfron M. Antiinflammatory activity and acute toxicity of *Dodonaea viscosa*. Fitoterapia. 2006;**77**(6):478-480. DOI: 10.1016/j.fitote.2006.06.002

[18] Getie M, Gebre-Mariam T, Rietz R, Höhne C, Huschka C, Schmidtke M, Neubert RHH. Evaluation of the anti-microbial and anti-inflammatory activities of the medicinal plants *Dodonaea viscosa, Rumex nervosus* and *Rumex abyssinicus*. Fitoterapia. 2003;**74**(1):139-143. DOI: 10.1016/S0367-326X(02)00315-5

[19] Salinas-Sánchez D, Herrera-Ruiz M, Pérez S, Jiménez-Ferrer E, Zamilpa A. Anti-inflammatory activity of hautriwaic acid isolated from *Dodonaea viscosa* leaves. Molecules. 2012;**17**(4):4292-4299. DOI: 10.3390/molecules17044292

[20] Salinas-Sánchez D, Zamilpa A, Pérez S, Herrera-Ruiz M, Tortoriello J, González-Cortazar M, Jiménez-Ferrer E. Effect of hautriwaic acid isolated from *Dodonaea viscosa* in a model of kaolin/carrageenan-induced monoarthritis. Planta Medica. 2015;**81**(14):1240-1247. DOI: 10.1055/s-0035-1546197

[21] Pathak A, Bhutani M, Nair A, Ahn K, Chakraborty A, Kadara H, Aggarwal B. Ursolic acid inhibits STAT3 activation pathway leading to suppression of proliferation and chemosensitization of human multiple myeloma cells. Molecular Cancer Research. 2007; **5**(9):943-955. DOI: 10.1158/1541-7786.MCR-06-0348

[22] Wand X, Zhang F, Yang L, Mei Y, Long H, Zhang X. Ursolic acid inhibits proliferation and induces apoptosis of cancer cells in vitro and in vivo. Journal of Biomedicine and Biotechnology. 2011;**2011**:1-8. DOI: 10.1155/2011/419343

[23] Woźniak Ł, Skąpska S, Marszałek K. Ursolic acid-a pentacyclic triterpenoid with a wide spectrum of pharmacological activities. Molecules. 2015;**20**(11):20614-20641. DOI: 10.3390/molecules201119721

[24] Vasconcelos M, Royo V, Ferreira D, Crotti A, Silva M, Carvalho J, Bastos JK, Cunha WR. In vivo analgesic and anti-inflammatory activities of ursolic acid and oleanoic acid from *Miconia albicans* (Melastomataceae). Zeitschrift für Naturforschung. 2006;**61**(7–8):477-482. DOI: 0939-5075/2006/0700-0477

[25] Baricevic D, Sosa S, Della-Loggia R, Tubaro A, Simonovska B, Krasna A, Zupancic A. Topical anti-inflammatory activity of *Salvia officinalis* L. leaves: The relevance of ursolic acid. Journal of Ethnopharmacology. 2001;**75**(2):125-132. DOI: 10.1016/S0378-8741(00)00396-2

[26] González-Chávez M, Ramos-Velázquez C, Serrano-Vega R, Pérez-González C, Sánchez-Mendoza E, Pérez-Gutiérrez S. Anti-inflammatory activity of standardized dichloromethane extract of *Salvia connivens* on macrophages stimulated by LPS. Pharmaceutical Biology. 2017; **55**(1):1467-1472. DOI: 10.1080/13880209.2017.1305423

[27] Zerin T, Lee M, Jang W, Nam K, Song H. Anti-inflammatory potential of ursolic acid in mycobacterium tuberculosis-sensitized and Concanavalin A-stimulated cells. Molecular Medicine Reports. 2016;**13**(3):2736-2744. DOI: 10.3892/mmr.2016.4840

[28] Formisano C, Sanna C, Ballero M, Chianese G, Sirignano C, Rigano D, Taglialatela-Scafati O. Anti-inflammatory sesquiterpene lactones from *Onopordum illyricum* L.(Asteraceae), an Italian medicinal plant. Fitoterapia. 2017;**116**:61-65. DOI: 10.1016/j.fitote.2016.11.006

[29] Youn UJ, Park EJ, Kondratyuk TP, Sripisut T, Laphookhieo S, Pezzuto JM, Chang LC. Anti-inflammatory triterpenes from the apical bud of *Gardenia sootepensis*. Fitoterapia. 2016;**114**:92-97. DOI: 10.1016/j.fitote.2016.08.012

[30] Garayev E, Herbette G, Di Giorgio C, Chiffolleau P, Roux D, Sallanon H, Baghdikian B. New sesquiterpene acid and inositol derivatives from *Inula montana* L. Fitoterapia. 2017; **120**:79-84. DOI: 10.1016/j.fitote.2017.05.011

[31] Mai NT, Cuc NT, Anh HLT, Nhiem NX, Tai BH, Yen PH, Oh H. Two new guaiane sesquiterpenes from *Datura metel* L. with anti-inflammatory activity. Phytochemistry Letters. 2017;**19**:231-236. DOI: 10.1016/j.phytol.2017.01.011

[32] Shen XY, Yu Y, Chen GD, Zhou H, Luo JF, Zuo YH, Dai Y. Six new sesquiterpenoids from *Nardostachys chinensis* Batal. Fitoterapia. 2017;**119**:75-82. DOI: 10.1016/j.fitote.2017.04.004

[33] Pan LL, Xu P, Luo XL, Wang LJ, Liu SY, Zhu YZ, Liu XH. Shizukaol B, an active sesquiterpene from *Chloranthus henryi*, attenuates LPS-induced inflammatory responses in BV2 microglial cells. Biomedicine & Pharmacotherapy. 2017;**88**:878-884. DOI: 10.1016/j.biopha.2017.01.152

[34] Wu ZN, Zhang YB, Chen NH, Li MJ, Li MM, Tang W, Wang GC. Sesquiterpene lactones from *Elephantopus mollis* and their anti-inflammatory activities. Phytochemistry. 2017;**137**: 81-86. DOI: 10.1016/j.phytochem.2017.01.020

[35] Shen CP, Luo JG, Yang MH, Kong LY. Sesquiterpene dimers from the roots of *Chloranthus holostegius* with moderate anti-inflammatory activity. Phytochemistry. 2017;**137**:117-122. DOI: 10.1016/j.phytochem.2017.02.016

[36] Van Kiem P, Hang DT, Nhiem NX, Tai BH, Anh HLT, Van Cuong P, Subedi L. Sesquiterpene derivatives from marine sponge *Smenospongia cerebriformis* and their anti-inflammatory activity. Bioorganic & Medicinal Chemistry Letters. 2017;**27**(7):1525-1529. DOI: 10.1016/j.bmcl.2017.02.040

[37] Fang X, Xu XK, Wang GW, Zeng RT, Tian XH, Shi ZR, Zhang WD. Guaianolide sesquiterpenoids from *Ainsliaea yunnanensis*. Phytochemistry. 2017;**139**:47-55. DOI: 10.1016/j.phytochem.2017.04.001

[38] Zhuo ZG, Wu GZ, Fang X, Tian XH, Dong HY, Xu XK, Shen YH. Chlorajaponols A–F, sesquiterpenoids from *Chloranthus japonicus* and their in vitro anti-inflammatory and anti-tumor activities. Fitoterapia. 2017;**119**:90-99. DOI: 10.1016/j.fitote.2017.04.009

[39] Walshe-Roussel B, Choueiri C, Saleem A, Asim M, Caal F, Cal V, Arnason JT. Potent anti-inflammatory activity of sesquiterpene lactones from *Neurolaena lobata* (L.) R. Br. ex Cass., a Q'eqchi'Maya traditional medicine. Phytochemistry. 2013;**92**:122-127. DOI: 10.1016/j.phytochem.2013.05.004

[40] de Albuquerque Ugoline BC, de Souza J, Ferrari FC, Ferraz-Filha ZS, Coelho GB, Saúde-Guimarães DA. The influence of seasonality on the content of goyazensolide and on anti-inflammatory and anti-hyperuricemic effects of the ethanolic extract of *Lychnophora passerina* (Brazilian arnica). Journal of Ethnopharmacology. 2017;**198**:444-450. DOI: 10.1016/j.jep.2017.01.017

[41] Liu Q, Jo YH, Kim SB, Jin Q, Hwang BY, Lee MK. Sesquiterpenes from the roots of *Lindera strychnifolia* with inhibitory effects on nitric oxide production in RAW 264.7 cells. Bioorganic & Medicinal Chemistry Letters. 2016;**26**(20):4950-4954. DOI: 10.1016/j.bmcl.2016.09.012

[42] Chi J, Li BC, Dai WF, Liu L, Zhang M. Highly oxidized sesquiterpenes from *Artemisia austro-yunnanensis*. Fitoterapia. 2016;**115**:182-188. DOI: 10.1016/j.fitote.2016.10.013

[43] Xia G, Zhou L, Ma J, Wang Y, Ding L, Zhao F, Qiu F. Sesquiterpenes from the essential oil of *Curcuma wenyujin* and their inhibitory effects on nitric oxide production. Fitoterapia. 2015;**103**:143-148. DOI: 10.1016/j.fitote.2015.03.021

[44] Hegazy MEF, Nakamura S, Tawfik WA, Abdel-Azim NS, Abdel-Lateff A, Matsuda H, Paré PW. Rare hydroperoxyl guaianolide sesquiterpenes from *Pulicaria undulata*. Phytochemistry Letters. 2015;**12**:177-181. DOI: 10.1016/j.phytol.2015.03.019

[45] Kim TH, Li H, Wu Q, Lee HJ, Ryu JH. A new labdane diterpenoid with anti-inflammatory activity from *Thuja orientalis*. Journal of Ethnopharmacology. 2013;**146**(3):760-767. DOI: 10.1016/j.jep.2013.02.001

[46] González Y, Doens D, Santamaría R, Ramos M, Restrepo CM, de Arruda LB, Fernández PL. A pseudopterane diterpene isolated from the octocoral *Pseudopterogorgia acerosa* inhibits the inflammatory response mediated by TLR-ligands and TNF-alpha in macrophages. PloS One. 2013;**8**(12):e84107. DOI: 10.1371/journal.pone.0084107.g003

[47] Wangchuk P, Navarro S, Shepherd C, Keller PA, Pyne SG, Loukas A. Diterpenoid alkaloids of *Aconitum laciniatum* and mitigation of inflammation by 14-O-acetylneoline in a murine model of ulcerative colitis. Scientific Reports. 2015;**5**:12845. DOI: 10.1038/srep12845

[48] Través PG, Pimentel-Santillana M, Rico D, Rodriguez N, Miethke T, Castrillo A, Boscá L. Anti-inflammatory actions of acanthoic acid-related diterpenes involve activation of the PI3K p110γ/δ subunits and inhibition of NF-κB. Chemistry & Biology. 2014;**21**(8):955-966. DOI: 10.1016/j.chembiol.2014.06.005

[49] Chen JJ, Ting CW, Wu YC, Hwang TL, Cheng MJ, Sung PJ, Chen JF. New labdane-type diterpenoids and anti-inflammatory constituents from *Hedychium coronarium*. International Journal of Molecular Sciences. 2013;**14**(7):13063-13077. DOI: 10.3390/ijms140713063

[50] Sun CP, Yuan T, Wang L, Kang N, Zhao F, Chen LX, Qiu F. Anti-inflammatory labdane-type diterpenoids from *Physalis angulate*. RSC Advances. 2016;**6**(80):76838-76847. DOI: 10.1039/c6ra16424b

[51] Premprasert C, Tewtrakul S, Plubrukarn A, Wungsintaweekul J. Anti-inflammatory activity of diterpenes from *Croton stellatopilosus* on LPS-induced RAW264. 7 cells. Journal of Natural Medicines. 2013;**67**(1):174-181. DOI: 10.1007/s11418-012-0668-5

[52] Cheng HH, Cheng YB, Hwang TL, Kuo YH, Chen CH, Shen YC. Randainins A–D, based on unique diterpenoid architectures, from *Callicarpa randaiensis*. Journal of Natural Products. 2015;**78**(8):1823-1828. DOI: 10.1021/acs.jnatprod.5b00012

[53] Lee M, Kim SH, Lee HK, Cho Y, Kang J, Sung SH. *ent*-kaurane and *ent*-pimarane diterpenes from *Siegesbeckia pubescens* inhibit lipopolysaccharide-induced nitric oxide production in BV2 microglia. Biological & Pharmaceutical Bulletin. 2013;**37**(1):152-157. DOI: 10.1248/bpb.b13-00233

[54] Qin J, Li HM, Gao LH, Liu D, Li RT, Lee KH. New labdane diterpenoids from *Leonurus japonicus* and their anti-inflammatory activity. Phytochemistry Letters. 2014;**10**:313-317. DOI: 10.1016/j.phytol.2014.10.027

[55] He YR, Shen YH, Shan L, Yang X, Wen B, Ye J, Zhang WD. Diterpenoid lanceolatins A–G from *Cephalotaxus lanceolata* and their anti-inflammatory and anti-tumor activities. RSC Advances. 2015;**5**(6):4126-4134. DOI: 10.1039/c4ra10665b

[56] Kapewangolo P, Omolo JJ, Bruwer R, Fonteh P, Meyer D. Antioxidant and anti-inflammatory activity of *Ocimum labiatum* extract and isolated labdane diterpenoid. Journal of Inflammation. 2015;**12**(1):4. DOI: 10.1186/s12950-015-0049-4

[57] Chen YC, Li YC, You BJ, Chang WT, Chao LK, Lo LC, Kuo YH. Diterpenoids with anti-inflammatory activity from the wood of *Cunninghamia konishii*. Molecules. 2013;**18**(1):682-689. DOI: 10.3390/molecules18010682

[58] Erharuyi O, Adhikari A, Falodun A, Jabeen A, Imad R, Ammad M, Gören N. Cytotoxic, anti-inflammatory, and leishmanicidal activities of diterpenes isolated from the roots of *Caesalpinia pulcherrima*. Planta Medica. 2016;**83**(01/02):104-110. DOI: 10.1016/j.bjp.2016.12.008

[59] Thao NP, Luyen BTT, Ngan NTT, Song SB, Cuong NX, Nam NH, Van Minh C. New anti-inflammatory cembranoid diterpenoids from the Vietnamese soft coral *Lobophytum crissum*. Bioorganic & Medicinal Chemistry Letters. 2014;**24**(1):228-232. DOI: 10.1016/j.bmcl.2013.11.033

[60] Chen H, Wang H, Yang B, Jin DQ, Yang S, Wang M, Guo Y. Diterpenes inhibiting NO production from *Euphorbia helioscopia*. Fitoterapia. 2014;**95**:133-138. DOI: 10.1016/j.fitote.2014.03.010

[61] Armah FA, Annan K, Mensah AY, Amponsah IK, Tocher DA, Habtemariam S. Erythroivorensin: A novel anti-inflammatory diterpene from the root-bark of *Erythrophleum ivorense* (a Chev.). Fitoterapia. 2015;**105**:37-42. DOI: 10.1016/j.fitote.2015.06.001

[62] Rohr M, Oleinikov K, Jung M, Sandjo LP, Opatz T, Erkel G. Anti-inflammatory tetraquinane diterpenoids from a *Crinipellis* species. Bioorganic & Medicinal Chemistry. 2017;**25**(2):514-522. DOI: 10.1016/j.bmc.2016.11.016

[63] Rossi A, Pace S, Tedesco F, Pagano E, Guerra G, Troisi F, Izzo AA. The hallucinogenic diterpene salvinorin A inhibits leukotriene synthesis in experimental models of inflammation. Pharmacological Research. 2016;**106**:64-71. DOI: 10.1016/j.phrs.2016.01.032

[64] Pierri EG, Castro RC, Vizioli EO, Ferreira CM, Cavalheiro AJ, Tininis AG, Santos AG. Anti-inflammatory action of ethanolic extract and clerodane diterpenes from *Casearia sylvestris*. Revista Brasileira de Farmacognosia. 2017;**27**:495-501. DOI: 10.1016/j.bjp.2016.12.008

[65] Mireku EA, Kusari S, Eckelmann D, Mensah AY, Tolantsi FM, Spteller M. Anti-inflammatory tirucallane triterpenes from *Anopyxis kaineana* Pierre (Engl.), (Rhizophoraceae). Fitoterapia. 2015;**106**:84-91. DOI: 10.1016/j.fitote.2015.08.007

[66] Romero-Estrada A, Maldonado-Magaña A, González-Christen J, Bahena SM, Garduño-Ramírez ML, Rodríguez-López V, Alvarez L. Anti-inflammatory and antioxidative effects of six pentacyclic triterpenes isolated from the Mexican copal resin of *Bursera copallifera*. BMC Complementary and Alternative Medicine. 2016;**16**(1):422-432. DOI: 10.1186/s12906 016-1397-1

[67] Zhou J, Wei XH, Chen FY, Li CJ, Yang JZ, Ma J, Zhang DM. Anti-inflammatory pentacyclic triterpenes from the stems of *Euonymus carnosus*. Fitoterapia. 2017;**118**:21-26. DOI: 10.1016/j.fitote.2017.01.015

[68] Chun SK, Lalita S, Joonseok O, Sun YK, Sang UC, Kang R. Bioactive triterpenoids from the twigs of *Chaenomeles sinensis*. Journal of Natural Products. 2017;**80**:1134-1140. DOI: 10.1021/acs.jnatprod.7b00111

[69] de Almeida PDO, Boleti APA, Rüdiger AL, Lourenço GA, da Veiga Junior VF, Lima ES. Anti-inflammatory activity of triterpenes isolated from *Protium paniculatum* oil-resins. Evidence-Based Complementary and Alternative Medicine. 2015, Article ID: 293768. DOI: 10.1155/2015/293768

[70] Huang J, Wang Y, Li C, Wang X, He X. Triterpenes isolated from acorns of *Quercus serrata* Var. Brevipetiolata exert anti-inflammatory activity. Industrial Crops and Products. 2016; **91**:302-309. DOI: 10.1016/j.indcrop.2016.07.033

[71] Killeen MJ, Linder M, Pontoniere P, Crea R. NF-κβ signaling and chronic inflammatory diseases: Exploring the potential of natural products to drive new therapeutic opportunities. Drug Discovery Today. 2014;**19**(4):373-378. DOI: 10.1016/j.drudis.2013.11.002

Chapter 5

Recent Developments in Selected Sesquiterpenes: Molecular Rearrangements, Biosynthesis, and Structural Relationship among Congeners

Shashikumar K. Paknikar and Kamlesh Pai Fondekar

Additional information is available at the end of the chapter

http://dx.doi.org/10.5772/intechopen.74998

Abstract

Recent developments in selected sesquiterpenoids are reviewed for the past one decade (2005–2017) with special reference to Mechanisms of multistep molecular rearrangements of some sesquiterpenes or derivatives based on isotopic labeling studies and extensive spectroscopic analysis such as molecular rearrangement of acetyl cedrene to cedrene follower, acid catalyzed rearrangement of moreliane-based triketone, synthesis of (−)-isocomene and (−)-triquinane by acid-catalyzed rearrangement of (−)-modhephene, Total synthesis of (+)-cymbodiacetal, BF_3 catalyzed molecular rearrangements of mono epoxides of α- and β-himachalenes, santonic acid: Zn-HCl-ether reduction. Insights into biosynthesis of albaflavenone, caryol-1(11)-ene-10-ol, (+)-koraiol, pogostol, patchouli alcohol and valerenadiene are discussed. Congeners for probing structure-biosynthetic relationship. This approach is discussed with the availability of very interesting results on the isolation of highly oxygenated secondary metabolites from endophytic fungi, *Xylaria* sp.

Keywords: molecular rearrangements, mechanisms, synthetic application, CCR, biosynthesis, labeling experiments, congeners

1. Introduction

Sesquiterpene carbon frameworks comprise the largest group of terpenoids or sometime referred as isoprenoids. Farnesyl diphosphate (FPP) having three olefinic linkages undergo cyclization to produce very large number major cyclic frameworks which are further modified by oxidative cleavages, molecular rearrangements, loss of carbon atoms. The aim of this chapter is to provide an overview of the recent developments in sesquiterpenes with particular reference to molecular rearrangements, biosynthesis and structural relationship among congeners. The coverage is not comprehensive but a focused review of the literature (2005–till September 2017) and only the relevant research articles having a link with the above areas are selected for discussion.

2. Mechanisms of multistep molecular rearrangements, insight into biosynthesis and congeners for probing structure-biosynthetic relationship of selected natural products

2.1. Molecular rearrangement of acetyl cedrene to cedrene follower

The acetylation of cedrene **1** can lead to various products depending on the reaction conditions. Paknikar et al. [1] undertook a detailed study on the acetylation of cedar wood oil (Virginia) with acetic anhydride and polyphosphoric acid in dichloromethane which leads, besides acetyl cedrene **2**, also to a minor product, 1,7,7-trimethyl-2,3-(3′4′-dimethylbenzo) bicyclo[3.2.1]-octane **3**, called the follower. Structural analysis of **3** (**Scheme 1**) shows that rings A, B, C of **2** are rearranged as B, A, C in follower **3**.

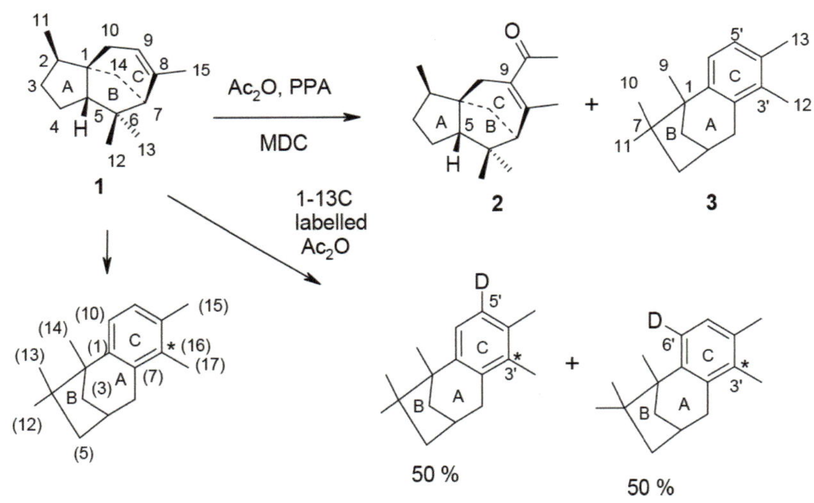

Scheme 1. Acetyl cedrene **2** and its follower **3**. The numbering in the brackets is the one from acetyl cedrene.

Formation of **3** from **2** can only be explained by a multistep intramolecular rearrangement. This shows that: (i) ring C of 2 has undergone initial ring enlargement and subsequent ring contraction; (ii) cleavage of the C6–C7 bond of 2 and formation of the new C6-C2 bond; (iii) enlargement of ring A of **2** with concomitant loss of water. The mechanism for the formation of **3** from **2** when 1-13C labeled acetic anhydride was used is shown in **Scheme 2**.

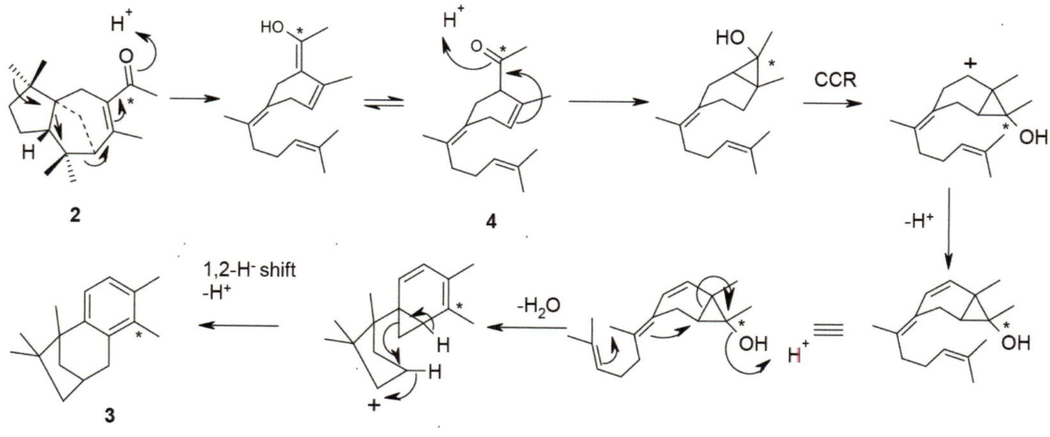

Scheme 2. Mechanism for the formation of follower **3** from acetyl cedrene **2**.

One characteristic feature of the formation of the follower **3** is sluggish reaction rates. Density Functional Theory (DFT) calculation of B3LYP/6-31G* type using the Gaussian version 09 (Gaussian) revealed that the first neutral intermediate **4** (**Scheme 2**) is higher in energy than acetyl cedrene by ~20 kcal. A series of further cascade- like cationic rearrangements is involved with breaking and bond-forming intermediates.

The formation of the neutral intermediate **4** is supported by the observation that this process is the reverse pathway for the biosynthesis of α-cedrene from FPP, which has been established previously [2]. Few other feasible mechanisms for the formation of follower **3** could be devised, and only the one presented fits the observation of 13C enriched label at the C-3′ position of follower **3**. Hence the key rearrangement is cyclopropylcarbinyl cation-cyclopropylcarbinyl cation rearrangement (CCR) [3, 4]. During the deuteriation of commercial acetyl cedrene, the follower was also deuterated, and it was observed that aromatic protons are exchanged. Interestingly, the product was only monodeuterated (**Scheme 1**) and the isotope was shared equally between the C-5′ and C-6′ positions of the follower **3**. This equal distribution of one deuterium atom between C-5′ and C-6′ can be accounted for by the facile 1,2-hydride and 1,2-deuteride shifts and equilibration.

2.2. Acid catalyzed rearrangement of moreliane based triketone. Characterization of keto lactone, a 1-11 seco-moreliane

An interesting molecular rearrangement has been reported by Morales and co-workers [5]. They observed that triketone **5** on treatment with *p*-TSA in benzene resulted in the formation of a keto lactone **6**, a 1–11 seco-moreliane derivative and also the first representative of this group (**Figure 1**).

The rearrangement depicted in **Scheme 3** involves initial cyclobutane ring expansion of the protonated triketone, generation of carbocationic intermediate **7** which rearranges *via* transition state in to protonated seco-moreliane **8**. These steps are supported by DFT calculations.

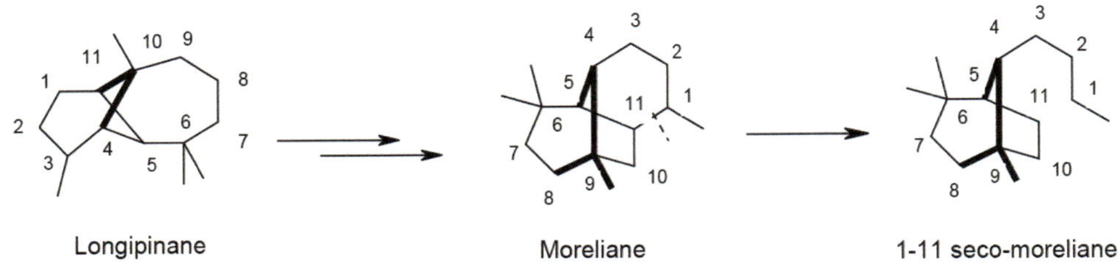

Figure 1. Skeletons of longipinane, moreliane and 1–11 seco-moreliane.

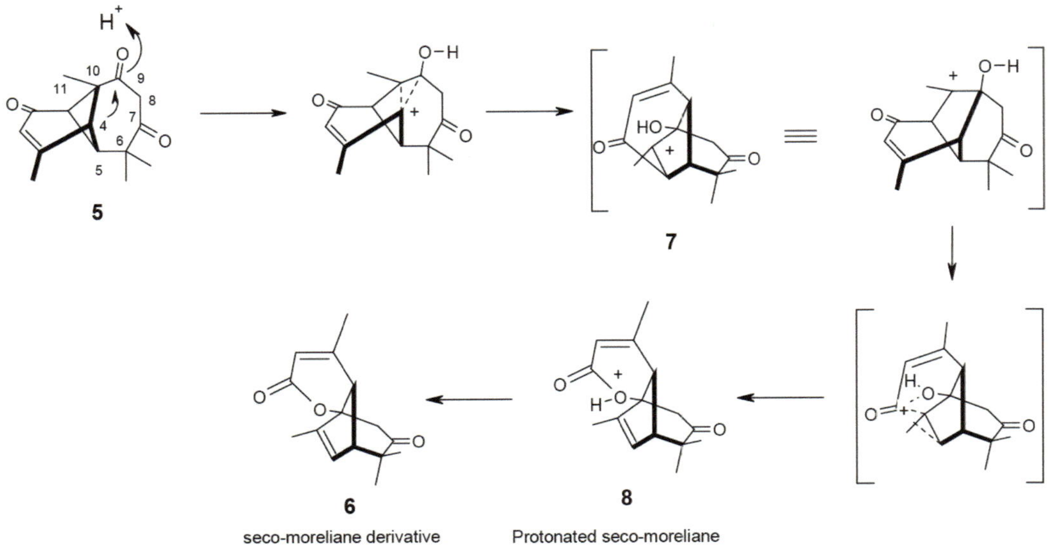

Scheme 3. Acid catalyzed rearrangement of triketone **5** to 1–11 seco-moreliane derivative **6**.

2.3. Synthesis of (−)-isocomene and (−)-triquinane by acid catalyzed rearrangement of (−)-modhephene

Triquinanes have received considerable attention by their unique structure as well as their reported biological activities. (−)-Modhephene **9** of established absolute stereochemistry was subjected to acid catalyzed carbocation rearrangements which led to an interesting synthesis of (−)-isocomene **10** and (−)-triquinane **11**[6]. This study was extended further by preparation of (−)-modhephene **9d** stereospecifically at 14β geminal methyl group. Under same experimental conditions, deuterium labeled (−)-triquinane **11d** a stereospecific 1,2-migration of 7/4β methyl group was observed (**Scheme 4**).

2.4. Total synthesis of (+)-cymbodiacetal

In 2010, Hayes and his co-workers reported [7] a total synthesis of (+)-Cymbodiacetal **12** by a biomimetic route proposed earlier [8, 9] using (R)-(+)-limonene **13**, the key step involves hetero Diels-Alder cycloaddition which proceeds with an *endo* selectivity (2:1) in a quantitative

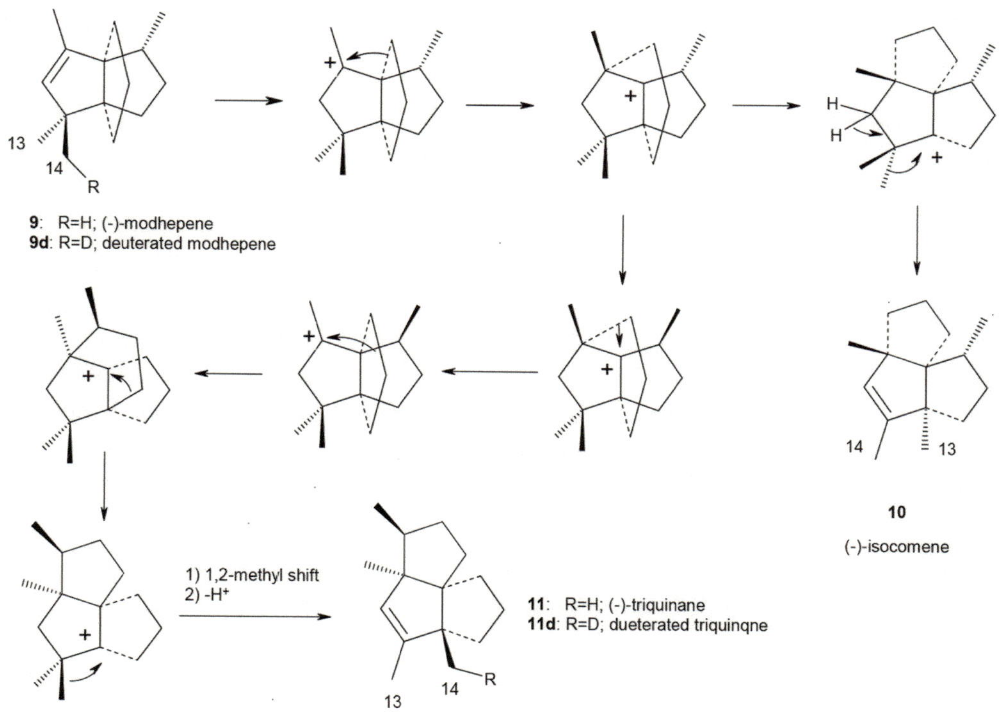

Scheme 4. Molecular rearrangement of (−)-modhephene **9** to (−)-isocomene **10** and (−)-triquinane **11**.

yield. Exploitation of *exo*-isomer with *m*-CPBA followed by acid catalyzed opening afforded (+)-cymbodiacetal **12** (**Scheme 5**). The uncertainty in absolute stereochemistry was independently established by X-ray crystallography. These studies also clarified discrepancies in the previously published work [8, 9].

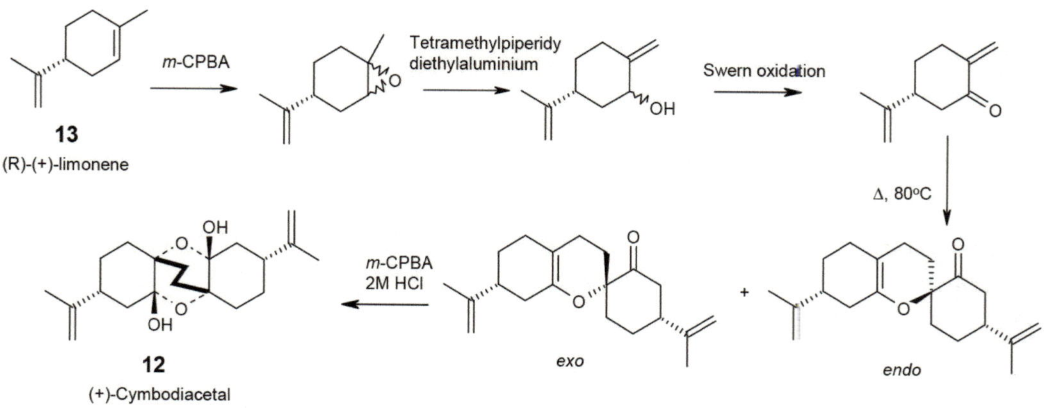

Scheme 5. Total synthesis of (+)-cymbodiacetal **12**.

2.5. BF_3 catalyzed molecular rearrangements of mono epoxides of α- and β-himachalenes

Previous examples of acid catalyzed rearrangements of sesquiterpenes have shown that the opening of the epoxide triggers the reaction and directs the subsequent molecular rearrangements. In practically, among all the cases the aim is to valorize the naturally occurring sesquiterpene hydrocarbons.

Manoury and co-workers [10] observed that on treatment of α-himachalene monoepoxide **14** with BF_3-Et_2O in CH_2Cl_2 at room temperature afforded a tricyclic ketone **16** (71% isolated yield) product along with an unsaturated alcohol **17** (18%). The structure **16** was unambiguously assigned to ketone based on ^1H, ^{13}C, ^1H-2D NMR experiments. The proposed mechanism (**Scheme 6**) involves ring opening of epoxide followed by participation of terminal methylene group to generate a tricyclic bridgehead carbocation **18** by ring contraction of seven membered ring to generate intermediate **19**. A stereospecific 1,4-hydride transfer is proposed in the last step to the formation of **16**.

Inspection of molecular models of intermediate **19** shows that the proposed stereospecific 1,4-hydride shift is unlikely and therefore a different process is responsible for the formation of ketone **16**.

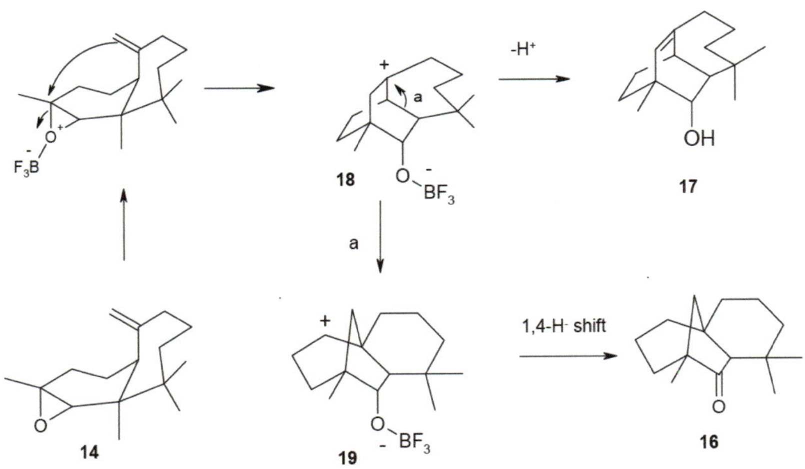

Scheme 6. Proposed mechanism for the formation of unsaturated alcohol **17** and tricyclic ketone **16**.

The structure assignment **17** to the minor product, a tricyclic unsaturated alcohol is based on spectral analysis and confirmed by single crystal X-ray data. The characteristic feature of **17** is the presence of a double bond involving a bridgehead carbon.

β-Himachalene monoepoxide **15** under identical experimental conditions gave two products major product (62%) and aryl-himachalene (10%). The major product was assigned structure **20**. The proposed mechanism explains formation of **20** (**Scheme 7**). The gross structure of this compound an allo-himachalol, a natural product isolated from *Cedrus deodara* [11].

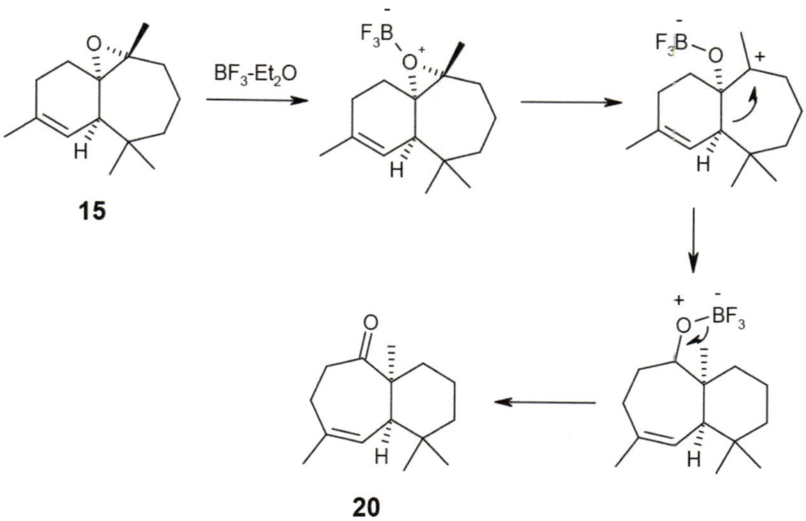

Scheme 7. Mechanism for BF_3 catalyzed transformation of β-himachalenes monoepoxide **15** to ketone **20**.

Compounds **16, 17** and **20** are all optically active and since the absolute stereochemistry of himachalenes are known, it is observed that C7 α-H of α-himachalenes remains intact throughout the rearrangement. The absolute stereochemistry of **16, 17** and **20** is shown in **Figure 2**.

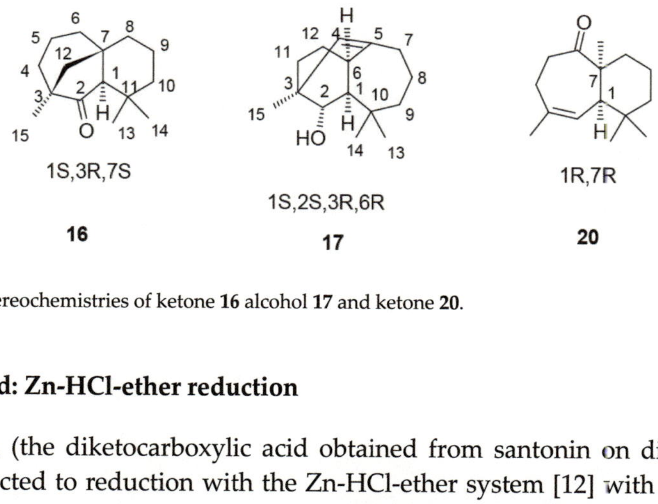

Figure 2. Absolute stereochemistries of ketone **16** alcohol **17** and ketone **20**.

2.6. Santonic acid: Zn-HCl-ether reduction

Santonic acid **21** (the diketocarboxylic acid obtained from santonin on digestion with aq. alkali) was subjected to reduction with the Zn-HCl-ether system [12] with an aim to obtain the previously prepared pinacol **22** *via* intramolecular pinacolisation primarily because of conformational structure of santonic acid with close proximity of the 1,4-diketone system. Under these conditions santonic acid **21** did not afford the pinacol **22**, but yielded a 60:40 mixture (GCMS, ^{1}H NMR) of succinic anhydride derivatives **23** and **24**. It is clear that the reaction proceeds *via* pinacol **22**, which, under strong acidic conditions, undergoes further rearrangement to give anhydrides **23** and **24** (**Scheme 8**).

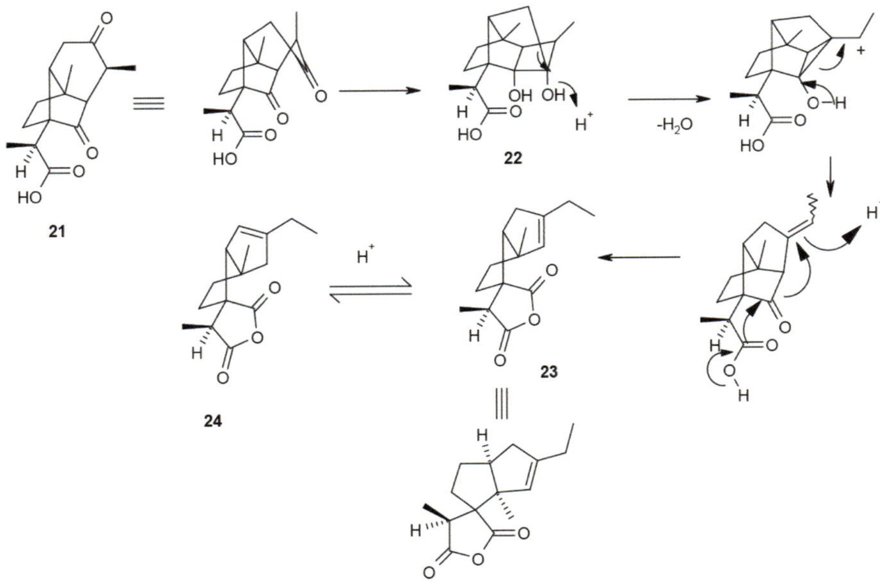

Scheme 8. Mechanistic pathway for the conversion of santonic acid **21** to bicyclo[3.3.0] octanes **23** and **24**.

2.7. Biosynthesis of albaflavenone

The tricyclic sesquiterpene antibiotic albaflavenone **25** isolated from the gram positive soil bacteria *Streptomyces coelicolor* A3 and *Streptomyces albidoflavus* is biosynthesized by enzymes encoded in a two-gene operm [13]. Initially, the sesquiterpene epi-isozizaene synthase catalyzes the cyclization of *2E, 6E*-farnesyl diphosphate (FPP) to (+)-epi-isozizaene **26**. A two-step allylic oxidation of **26** catalyzed by a single cytochrome P450170A1 (crP170A1) results in the formation of (+)-albaflavenone **25** *via* an epimeric mixture of (*5S*)-albaflavenol **27** and (*5R*)-albaflavenol **28** intermediates (**Scheme 9**) [14].

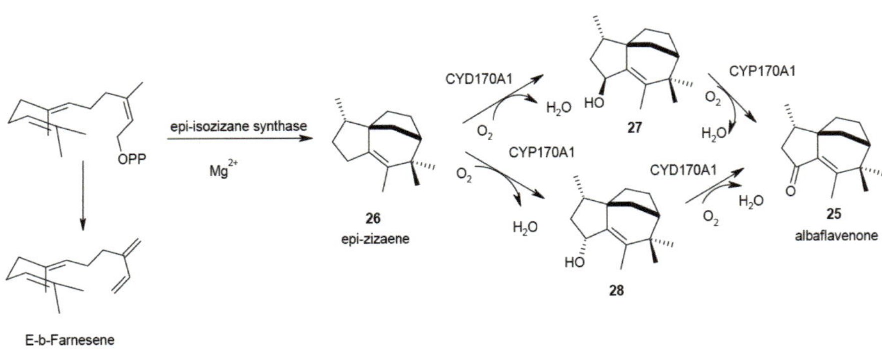

Scheme 9. Biosynthetic pathway of albaflavenone **25**.

The mechanism and stereochemistry of FPP to epi-isozizaene **26** *via* (*3R*)-nerolidyl diphosphate **29** has been conclusively established by labeling studies [15]. The entire biosynthetic process from FPP to epi-isozizaene is shown (**Scheme 10**). A two-step chemical synthesis of albaflavenone **25** from epi-isozizaene **26** was reported in this study.

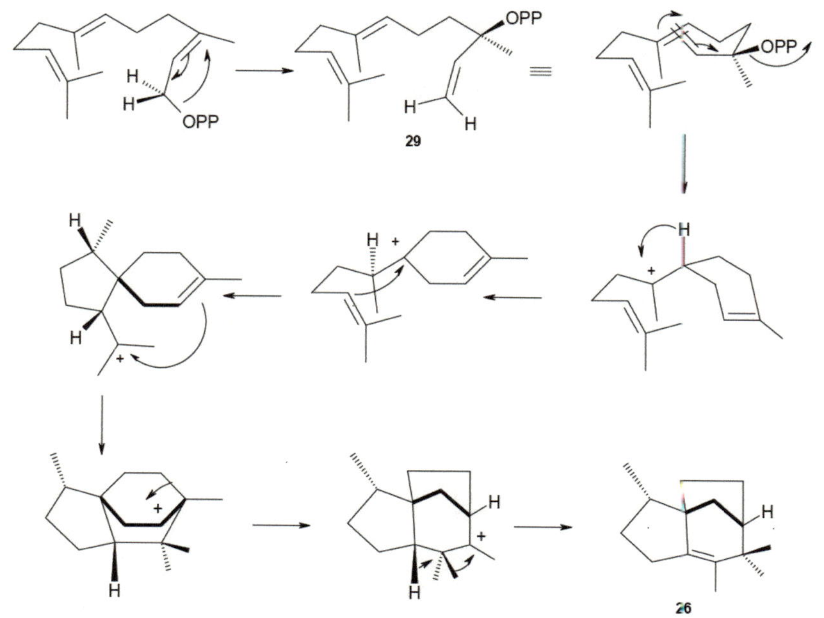

Scheme 10. Mechanism of the cyclization of *E,E*-FPP to epi-isozizaene **26** *via* (3*R*)-nerodilyl diphosphate **29**.

Ito and co-workers [16] reported a concise nine step total synthesis of albaflavenone without use of any protecting groups. Moreover, the absolute configuration of naturally occurring (+)-albaflavenone has been unambiguously established as *1S*, *7S* and *8R*.

2.8. The biosynthesis of caryol-1(11)-ene-10-ol: on the mechanism of the formation of caryolene: a putative biosynthetic precursor to caryol-1(11)-ene-10-ol

In 2013, Nguyen and Tantillo [17] investigated the mechanism of the formation of caryolene **30**, a putative biosynthetic precursor to caryol-1(11)-ene-10-ol **31** by DFT calculations (**Figure 3**).

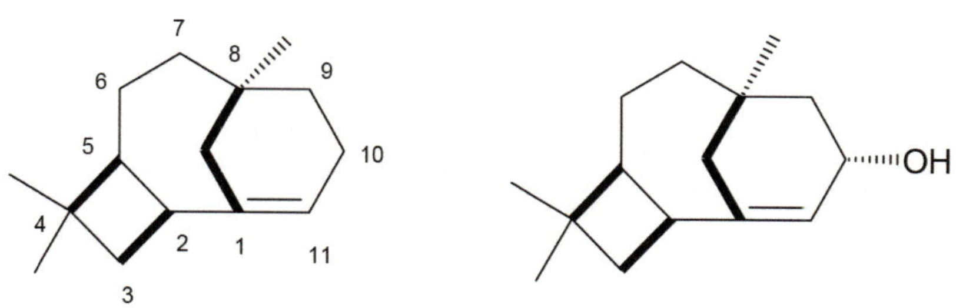

30 **31**

Figure 3. Structures of caryolene **30** and caryol-1(11)-en-10-ol **31**.

Quantum chemical calculations indicated the mechanism involving a secondary carbocation intermediate **32** is not energetically viable. They proposed two mechanisms for caryolene **30** formation (pathway a and b). The pathway involves a base catalyzed deprotonation/reprotonation sequence and a tertiary carbocation minima (more likely) whereas pathway b involves intramolecular proton transfer and the generation of a secondary carbocation minima. Both mechanisms are predicted to involve concerted suprafacial/suprafacial [2 + 2] cycloaddition, whose asynchomicity allows them to avoid the constrains of orbital symmetry (**Scheme 11**).

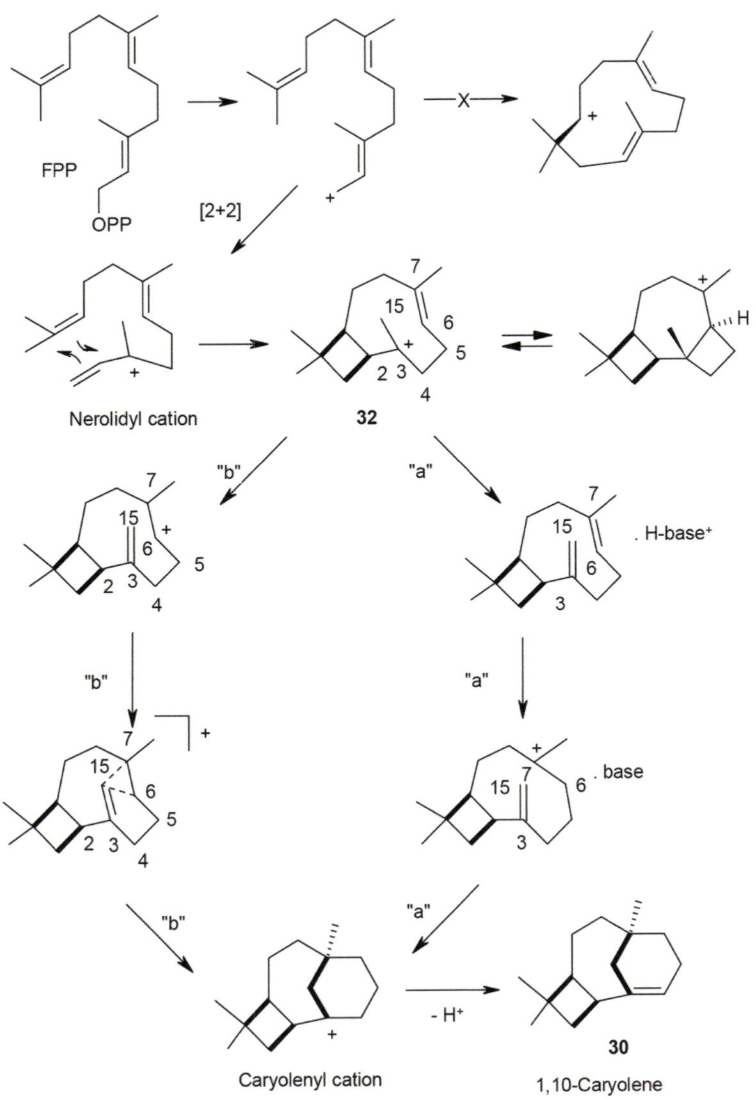

Scheme 11. Proposed mechanisms for the formation of 1,10-caryolene **30**.

2.9. Biosynthesis of (+)-koraiol

As an outcome of Tantillo's mechanism for caryolene **30** [17], biosynthetic pathway for koraiol **31** becomes evident (**Scheme 12**).

http://dx.doi.org/10.5772/intechopen.74998

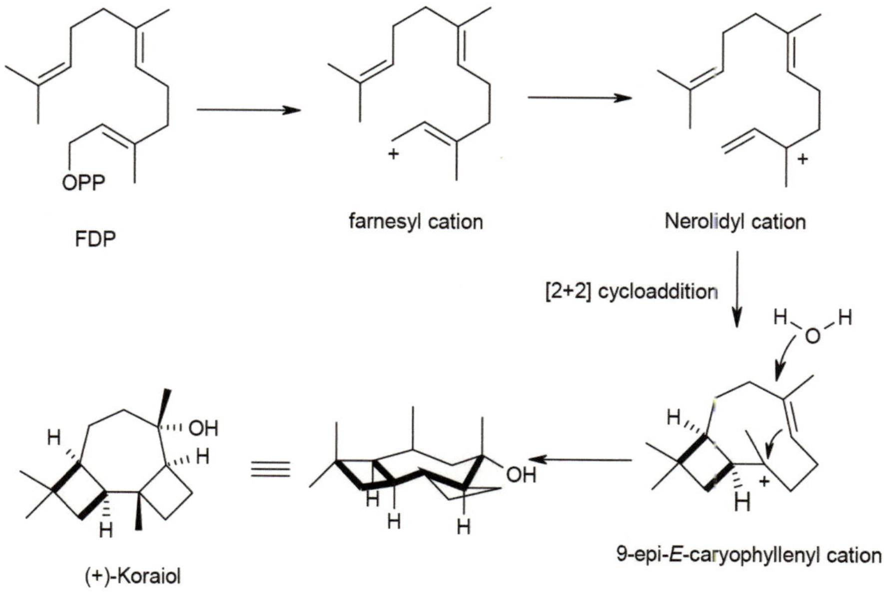

Scheme 12. Biosynthesis of (+)-koraiol **31**.

9-epi-*E*-Caryophyllene **32**, caryophyllene **33** and (+)-koraiol **31** were identified by Dickschat and co-workers [18, 19] who carried out investigation on the volatiles of *Fusarium fujikuroi* by the use of CLSA-GCMS. The sesquiterpenoids were divided in to two groups based on their proposed biosynthetic pathways. Volatile sesquiterpenoids produced by sesquiterpene cyclase Ffsc4 were characterized as β-caryophyllene and an optically active alcohol (+)-koraiol **31**. The structure **31** was assigned by extensive spectral analysis. The relative configuration of (+)-koraiol was elucidated by NOESY experiments. The *cis* fusion of rings A and B was deduced from the NOESY couplings of the bridge head hydrogen atoms 1H and 9H with each other with methyl protons 15-H and the pro-5-methylene protons 3-H. Interestingly, Khan et al. isolated (+)-koraiol, $[\alpha]_D + 31.7°$ from the oleoresin of Korean pine (*Pinus koraiensis* Sieb.). The relative stereochemistry as shown in **31** has been established by X-ray analysis [20]. The absolute stereostructure of the rare sesquiterpene (+)-9-epi-*E*-caryophyllene, an enantiomer of **32** was isolated from *Dacrydium cupressinum* by Weavers and co-workers [21] (**Figure 4**).

It is tempting to speculate (+)-koraiol **31** is biosynthesized from 9-epi-*E*-caryophyllene **32**.

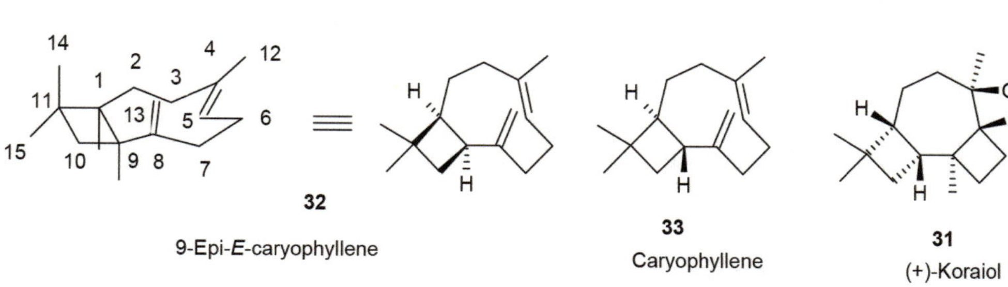

Figure 4. Structures of 9-epi-*E*-Caryophyllene **32**, caryophyllene **33** and (+)-koraiol **31**.

2.10. Biosynthesis of Pogostol

Biosynthesis of pogostol **34** by the endophytic fungus *Geniculosporium* was investigated by Dickschat and co-workers [22]. In this study, six 13C labeled isotopomers of mevalonolactone were synthesized and used in feeding experiments with the endophytic fungus *Geniarlosperium*. Feeding experiments with **35a** and **35b** gave insights into the stereochemical course of the terpene cyclization. The methyl group of the mevalonolactone that is labeled in these two isotopomers is converted into terminal (z)-methyl group of FPP (C-13). Both feeding experiments showed that the deprotonation step leading to germacrene A **36** proceeds with stereospecific deprotonation of C-13 and not C-12 of FPP (**Figure 5**).

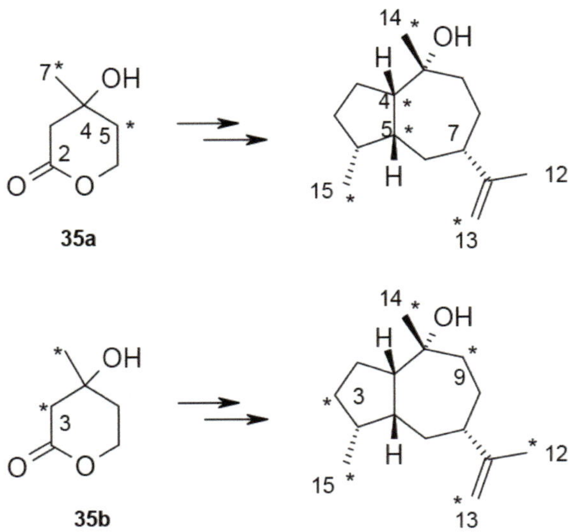

35a

35b

Figure 5. Biosynthesis of Pogostol **34** using isotopomers of mevalonolactone.

The volatile fraction was extracted by closed loop stripping apparatus followed by direct ^{13}CNMR analysis (CLSA-NMR) newly developed by the same group. The biosynthesis of pogostol **34** proceeds through initial formation of germacrene-A **36**. Protonation of 4,5 double bond initiates a second cyclization to cation which gets neutralized with water to give pogostol **34** (**Scheme 13**).

In view of correlation of (−)-pogostol **37** with (+)-bulnesol **38** with known absolute stereochemistry, (−)-pogostol be represented by the stereostructure **37** [23–25]. The stereostructure **34** thus represents (+)-pogostol (**Figure 6**).

2.11. Biosynthesis of patchouli alcohol (patchoulol)

The history of patchouli alcohol **39** from its isolation till date has narrated in a recent exhaustive review article [26]. Biosynthetic pathways were proposed based on experimental work for the conversion of FPP to patchouli alcohol **39** (**Scheme 14**).

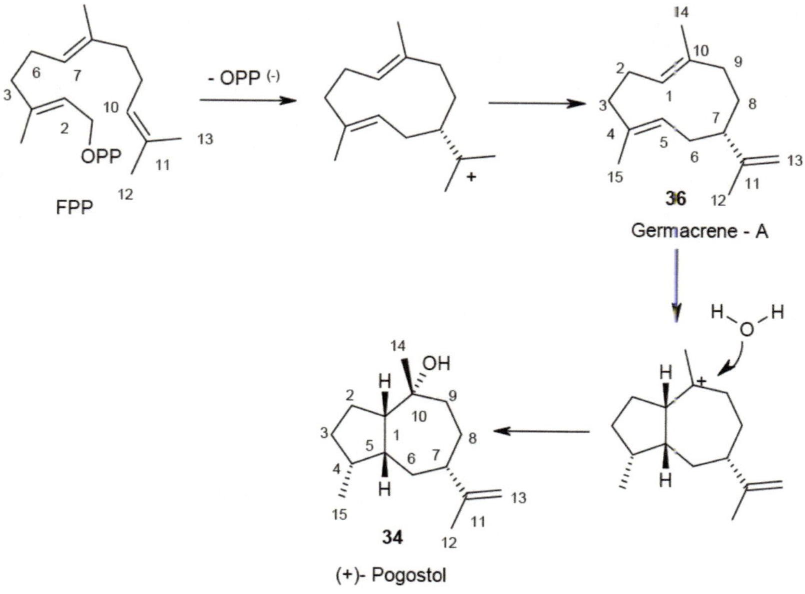

Scheme 13. Mechanism of pogostol **34** formation from FPP.

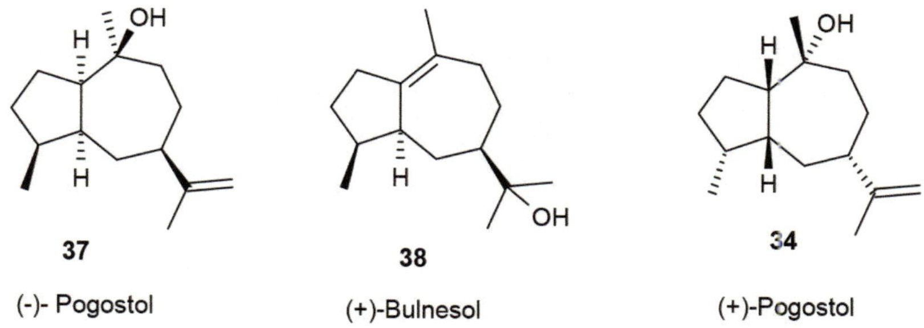

Figure 6. Absolute stereochemistry of (−)-pogostol **37**—correlation of (−)-pogostol **37** and (+)-bulnesol **38**.

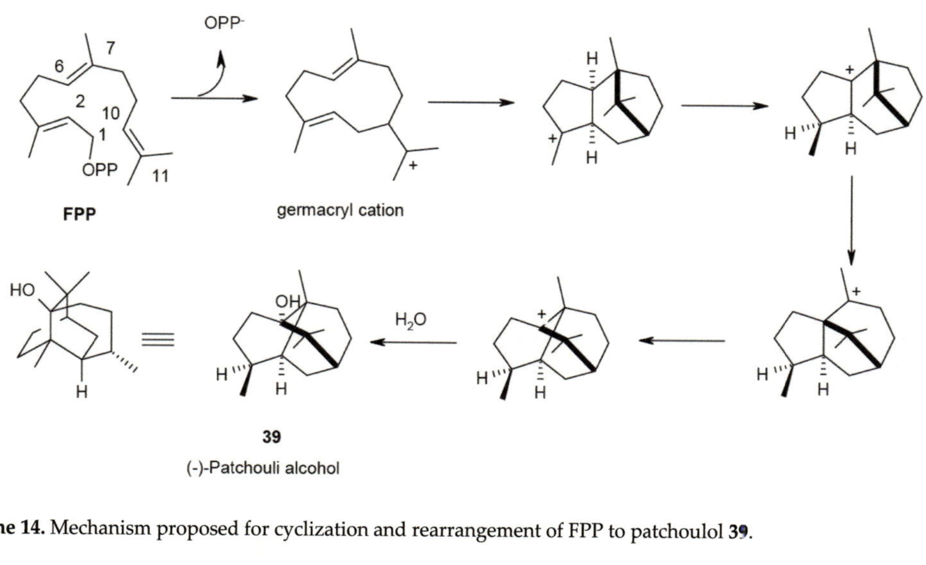

Scheme 14. Mechanism proposed for cyclization and rearrangement of FPP to patchoulol **39**.

Croteau et al. [27] and Akhila et al. [28] proposed biosynthetic pathways for the conversion of FPP to patchouli alcohol **39** based on experimental work. Croteau et al. reported the 1,3-shift for conversion of **40** to **41** while Akhila et al. proposed two consecutive 1,2-hydride shifts for the same conversion (**Scheme 15**).

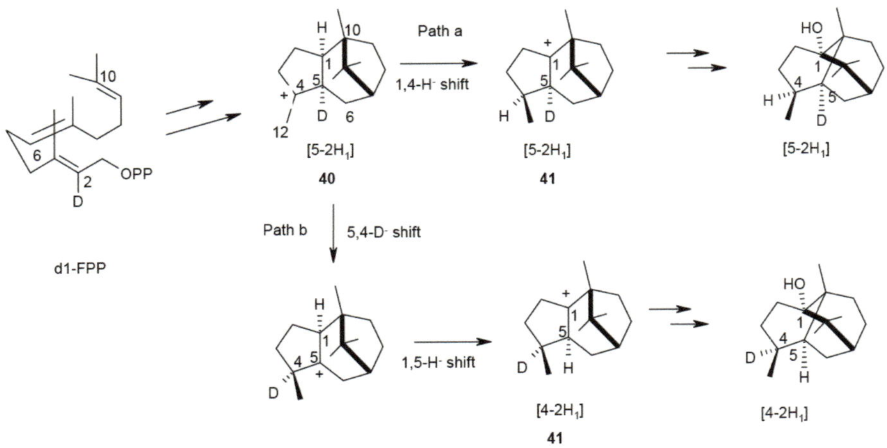

Scheme 15. Biosynthetic pathways for the conversion of [2-2H_1]-FPP to patchoulol isotopomer.

The recent isotopic labeling studies of Coates and colleagues [29] unrevealed the biosynthetic pathways for **39** which confirmed the 1,3-hydride shift across the five membered ring ruling out two consecutive 1,2-hydride shifts (**Scheme 16**).

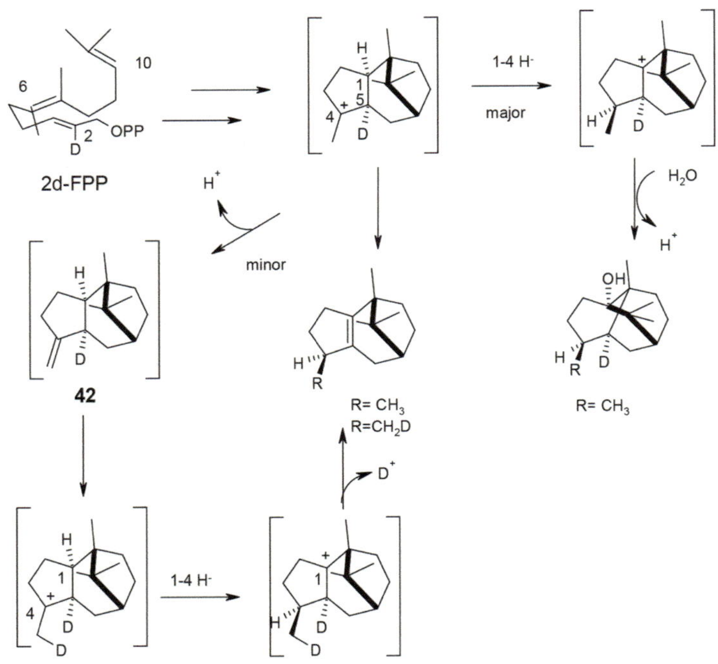

Scheme 16. Proposed biosynthesis of patchouliol **39** from deuterated FPP.

Incubation of isotopically pure [2-2H$_1$] (*E,E*)-farnesyldisulfate with recombinant patchoulol synthase (rPTS) from *Pogostemon cablin* afforded a 65:35 mixture of monodeuterated and di-dueterated patchouliols and several hydrocarbons of which eight have been identified. This is confirmed by extensive NMR analysis on the labeled patchouliol mixture and comparison with those of unlabeled patchouliol. Deuterium label was located at position C5 (both isotopomers ca. 100%) and at C12 (minor isotopomer, 30–35%). The formation of [5,12-2H$_2$] patchouliol is rationalized through an unknown (so far) hydrocarbon **42** which could incorporate deuterium at C12. This significant observation may have implication on the biosynthesis of nor-patchouliol **43** a congener of patchouliol, the biosynthesis is based on the earlier work [26] (**Figure 7**).

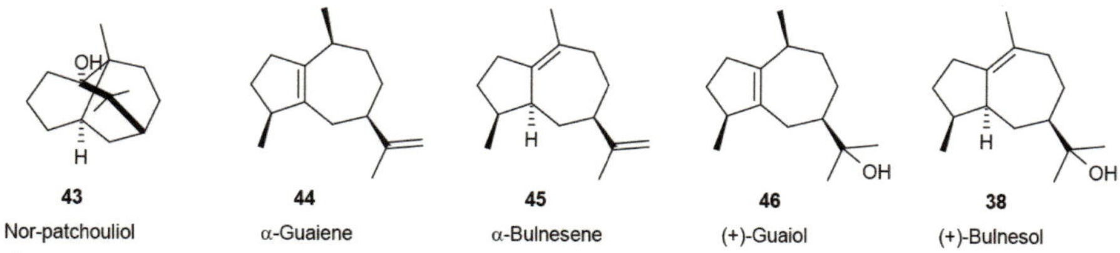

Figure 7. Structures of nor-patchouliol **43**, α-guaiene 44, α-bulnesene **45**, (+)-guaiol **46** and (+)-bulnesol **38**.

The interesting observation which can be made on the patchouli oil constituents that though α-guaine **44** and α-bulnesene **45** are genuine natural products [26], (+)-guaiol **46** and (+)-bulnesol **38** has never been reported to be present in patchouli oil.

2.12. Biosynthesis of Valerenadiene

Pyle et al. [30] reported the first enzymatic synthesis of valerena-4,7(11)-diene **47** (numbering used for valarenic acid) by a unique TPS from *Valeriana officinalis*. They identified two TPS's VoTPS1 and VoTPS2. Transgenic yeast expressing VoTPS1 produced germacrene B **48**, germacrene C **49** and germacrene D **50**. On the other hand, VoTPS 2 produced valerena-4,7(11)-diene **47** as a major compound was substantiated by ^{13}CNMR and GC–MS comparison with the synthetic standard. Minor products were identified as bicyclogermacrene **51** and alloaromadendrene **52**. The proposed mechanism involves ring contraction of germacrane ring to a nine-membered intermediate having isobutenyl side chain. Cyclization gives valerena-4,7(11)-diene **47** (**Scheme 17**).

Yeo et al. [31] proposed a mechanism wherein the isobutyl side chain is derived by the intermediacy of a caryophyllenyl carbocation **53**. A 1,2-hydride shift followed by opening of the cyclobutyl ring. In this way the two methylene carbons of the isobutenyl side chain are predicted to arise from C1 and C11 of the originating FPP and therefore should become labeled when [1-13C] acetate is incorporated into FPP by mevalonate pathway operating in yeast (**Scheme 18**).

Valerina-1-10-diene **47** and related sesquiterpenes retain an isobutyl side chain whose origin has been recognized as enigmatic because a chemical rationalization for their biosynthesis has not been obvious. They identified seven *Valeriana officinalis*, terpene synthase genes (VoTPSs) and two were functionally characterized as sesquiterpene synthase VoTPS1 and

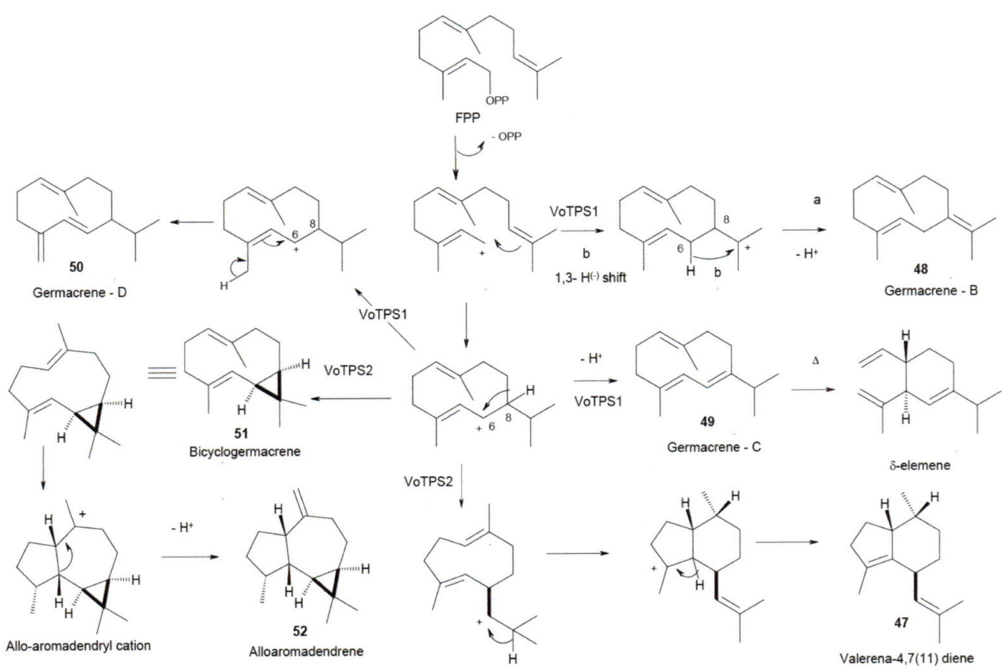

Scheme 17. Biosynthesis pathway for valerena-4,7(11) diene **47** and other sesquiterpenes from VoTPS1 and VoTPS2.

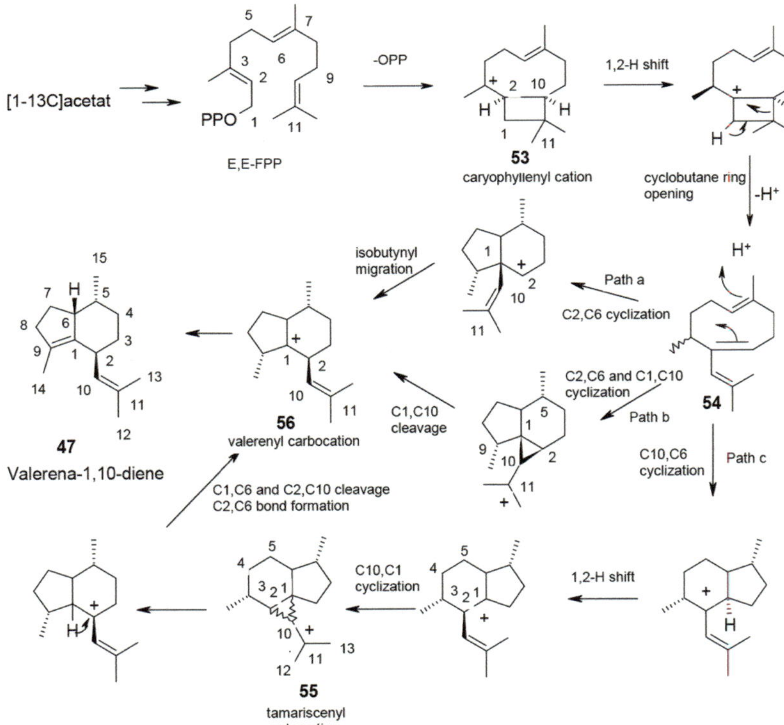

Scheme 18. Three biosynthetic pathways for valerena-4,7(11) diene **47** and other sesquiterpenes from VoTPS1.

VoTPS7. VoTPS7 encodes for a synthase that biosynthesizes germacrene C **49** (90%) whereas VoTPS 1 catalyzes conversion of *E,E*-FPP to valerena-1-10-diene **47**. Overexpression of VoTPS produced valarena-1-10-diene **47** on the basis of one and two dimensional NMR analysis, further confirmed by comparison with published spectral data, GC retention time and EIMS fragmentation pattern. The most characteristic feature of the [1-13C] acetate is the FPP derived from the incorporation of [1-13C] acetate had labels located at C1, C3, C5, C7, C9 and C11 as expected using a yeast expression system, specific labeled [1-13C] acetate. FPP was catalytically cyclized (using VoTPS1) and produce valeriana-1,10-diene **47** whose 13C labels were found at C3, C5, C7, C9, C1 and C11. Of these C1 and C11 were adjacent carbons of the isobutyl side chain. The proposed mechanism involves an intermediate of a caryophyllenyl carbocation **53**, 1,2-hydride shift followed by cleavage of C10-C11 bond generates a neutral monocyclic triene **54**. The proposed scheme also indicates formation of other sesquiterpenes through intermediates tamariscenyl cation **55** and valerenyl cation **56**.

Based on the experimental labeling data of Pyle et al. [30] and Yeo et al. [31], Paknikar et al. [4] proposed a new alternate biosynthetic route (**Scheme 19**) from IPP to valerenadiene **47** which fits the unusual 13C labeling found in valerian and avoids the previously unreported triene **54**.

In **Scheme 19**, the 2-1-10-11 sequence of carbons in the first cyclic intermediate **57** from *E,E*-FPP becomes 2-10-1-11 in valerenadiene **47** which fits the 13C labeling pattern formed from [1-13C] acetate [4]. The biosynthetic pathway involves one neutral intermediate; bicyclogermacrene **36** found in valerian [32]. The key reaction is a cyclopropylcarbinyl cation-cyclopropylcarbinyl cation rearrangement (CCR) analogues to a key reaction in the biosynthesis of squalane from resqualene [3]. Structure interrelationships of the congeners of valerenadiene **47** including bicyclogermacrene **36**, aromadendrene **51**, germacrene C **49**, germacrene D **50**, α-gurjunene **58** and malliol **59** were considered in this alternate pathway.

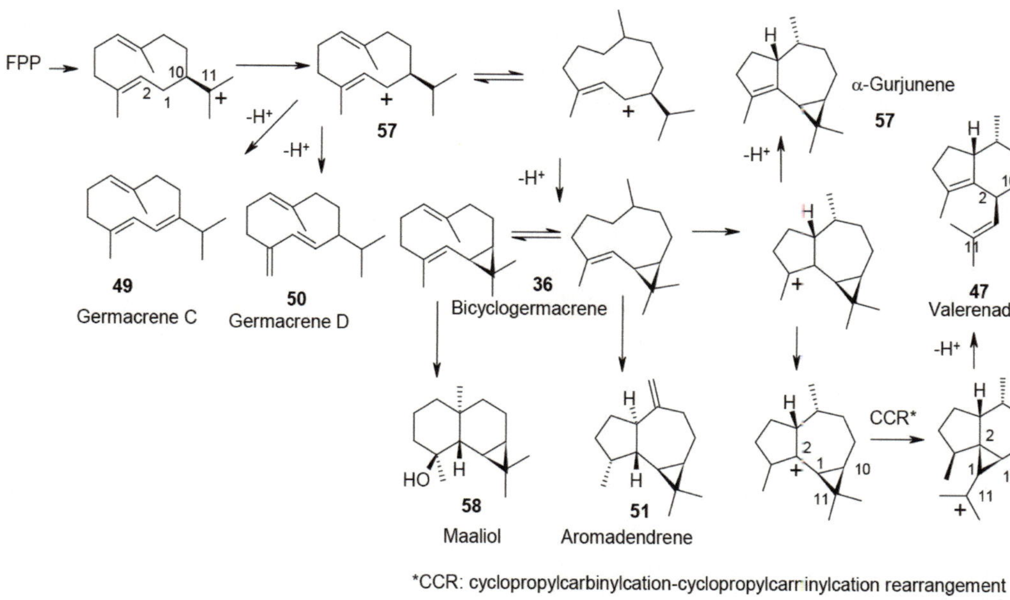

Scheme 19. A cyclopropropane route to valerenadiene **47** (numbering based on FPP).

Bicyclogermacrene **36** appears also to be an intermediate in the biosynthesis of related set of sesquiterpene with different stereochemistry found in *Valeriana officinalis*, including tamariscene **60**, pacifigorgiol **61** and (+)-pacifigorgia-1,10-diene **62** (**Scheme 20**). In this scheme also the key reaction is again cyclopropylcarbinyl cation-cyclopropylcarbinyl cation rearrangement (CCR) with this time with a different stereoisomer.

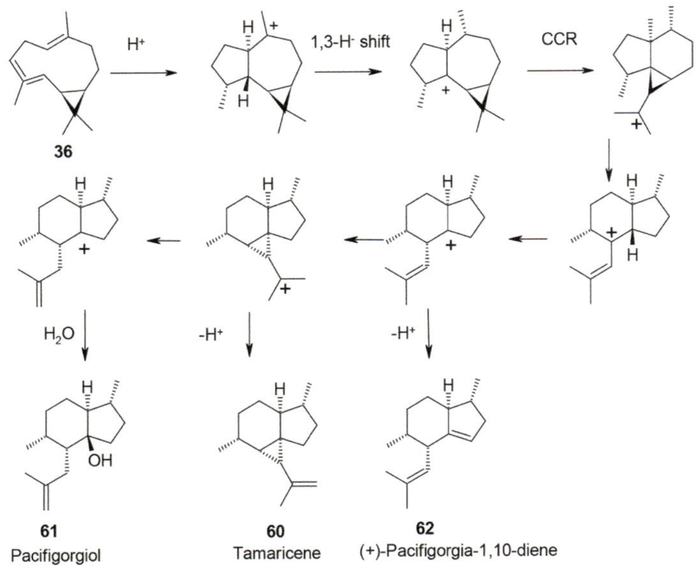

Scheme 20. Biosynthetic pathway of tamariscene **60**, pacifigorgiol **61** and (+)-pacifigorgia-1,10-diene **62** from bicyclogermacrene **36**.

Based on the results of three groups [4, 30, 31] a new consolidated mechanism for the biosynthesis of valerenadiene **47** from FPP *via* bicyclogermacrene **36** through alloaromadendryl cation **63** and CCR is presented which also explains formation of alloaromadendrene **64** (**Scheme 21**) replace alloaromadendryl cation with allo-aromadendryl cation.

2.13. Congeners of *Xylaria* sp.: structural interrelations

Endophytic fungi are reported to produce a number of bioactive metabolites and serve as an excellent source of highly oxygenated compounds which are likely to be potential drugs and also for the applications in crop science. The fungi belonging to genus *Xylaria* produces plethora of biologically related and structurally fascinating cadinenic and eudesmanic sesquiterpenes.

Liu and coworkers [33] reported isolation of highly oxygenated cadinane based compounds, three new xylaric acid A **65**, xylaric acid B **66** and xylaric acid C **67** and nine known compounds xylaric acid D **68**, heptelidic acid (avocetlin) [34] **69** hydroheptelidic acid **70**, gliocladic acid **71**, chlorheptelidic acid **72**, trichoderonic acid A **73**. The structure assignments are based on extensive spectral analysis. All these congeners belong to cadinane or seco-cadinane group of sesquiterpenes (**Figure 8**). The stereochemistry at C6 and C7 is unchanged for all the metabolites where C1 remains same for **66, 67, 69, 70, 72** and **73** and changes for **65, 68** and **71**.

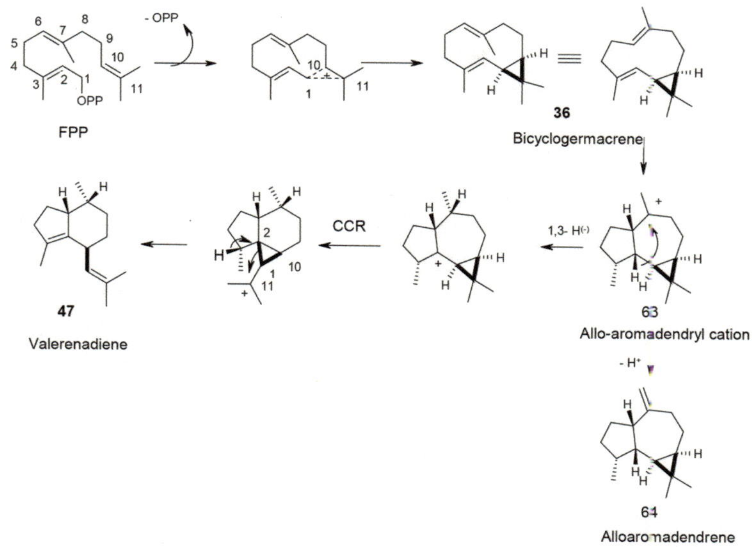

Scheme 21. Proposed new consolidated mechanism for the biosynthesis of valerenadiene 47.

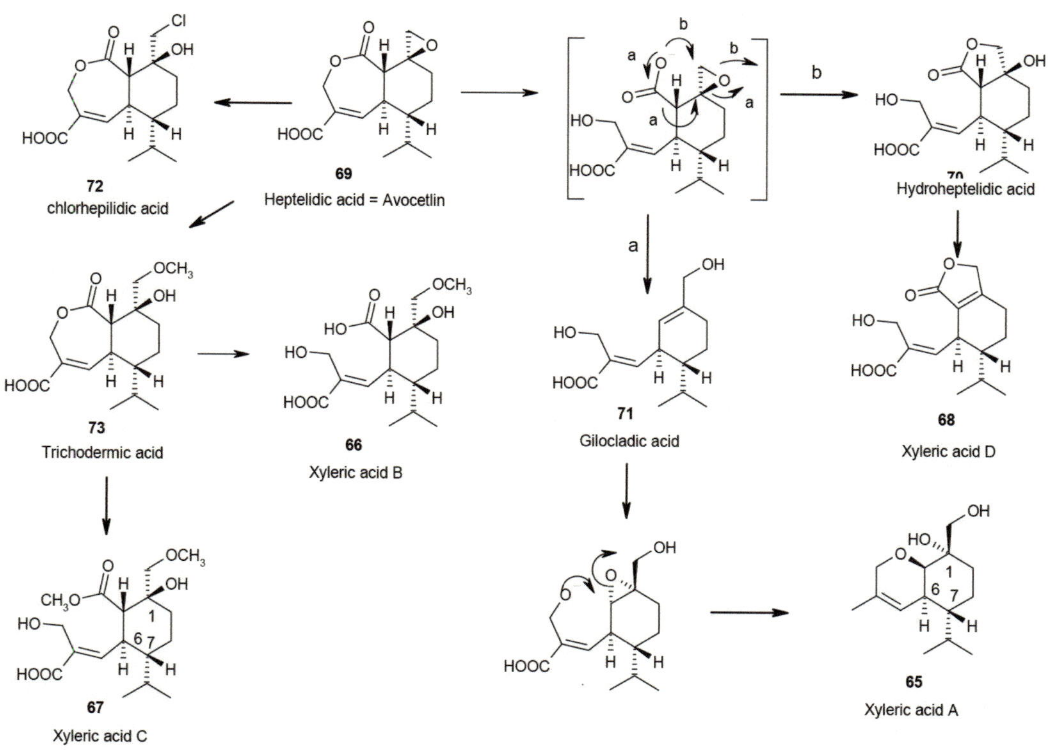

Figure 8. Structural interrelations among the congeners of *Xylaria* sp. and the sequence of formation of isolated metabolites 65–73.

Knowing the absolute stereochemistry of the congeners and their fungal origin, they belong to the "antipodal" set of compounds and they can be regarded as a result of extensive oxidative reactions of (−)-γ-cadinene **74**. Recently, Rabe *et al.* [35] have reported isolation of several sesquiterpenes including (−)-γ-cadinene, $[\alpha]_D$-32.3° by incubation of FPP with six purified bacterial terpene cyclases. The results were further supported by labeling experiments with 13C labeled isotopomers of FPP. Interestingly, antipodal cadinenic sesquiterpenes with known absolute configurations have been isolated from Indian vetiver oil (*Vetiveria zizanioides*) [36]. Isolation of (−)-γ-cadinene **74**, khusinol **75** and khusinol oxide **76** could be regarded as the precursors for the metabolites of *Xylaria* sp. A very clean sequence indicating a plausible order of formation of *Xylaria sp.* metabolites associated with the termite nest is presented (**Scheme 22**). We believe that this presentation will be useful while investigating the biosynthetic pathways using isotopic labeling studies.

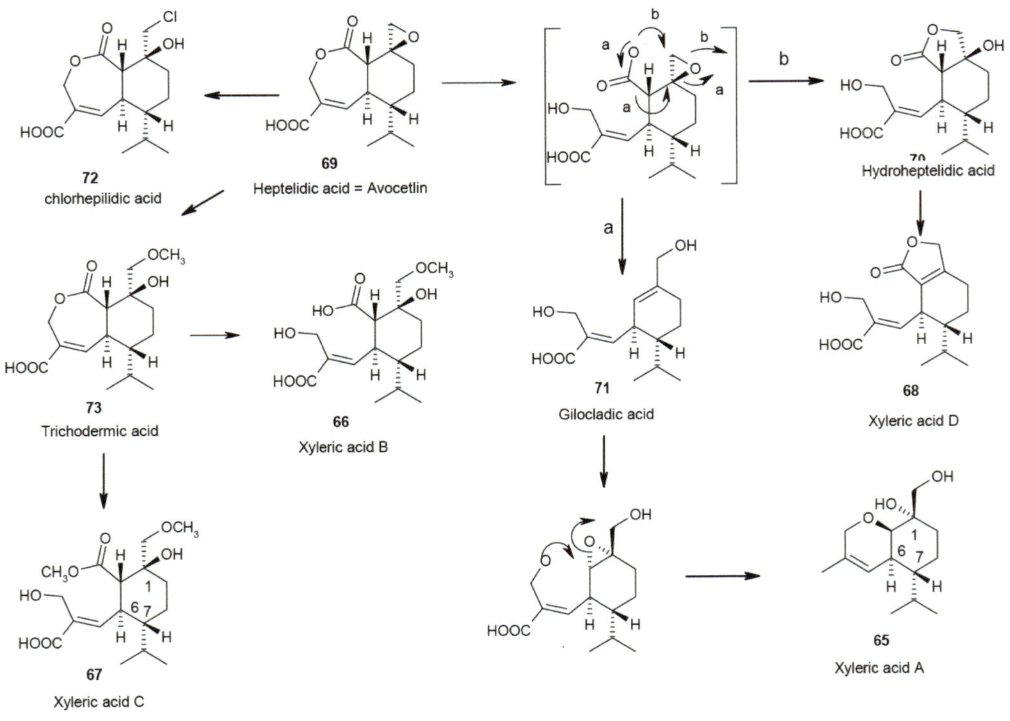

Scheme 22. Proposed plausible order of formation of *Xylaria* sp. metabolites from FPP.

3. Conclusions

This chapter gives overview of some of the interesting molecular rearrangements of sesquiterpenes reported over last decade. Further biosynthesis of albaflavenone, caryol-1(11)-ene-10-ol, (+)-koraiol, pogostol, patchouli alcohol and valerenadiene are also presented. The recent trends in the biosynthesis of natural products is focused on enzymatic synthesis using isotopic labeling, nevertheless discussions on structural interrelationships of various congeners provides insights in to natural occurrence of these molecules and finding their biosynthetic links.

Acknowledgements

We wish to dedicate this review to Professor R. B. Bates on his retirement from Research. We thank Dr. Asha D'Souza, Prof. Shailesh Shah and Rahul Chowgule for their valuable help in providing many research articles required for this review.

Author details

Shashikumar K. Paknikar[1]* and Kamlesh Pai Fondekar[2]

*Address all correspondence to: skpakni@yahoo.co.in

1 Siddharth Chemicals, Kundaim Industrial Estate, Kundaim, Goa, India

2 Deccan Fine Chemicals, Technology and Engineering, Santa Monica Works, Corlim, Goa, India

References

[1] Paknikar SK, Kamounah FS, Hansen PE, Wadia MS. Multi-step rearrangement mechanism for acetyl cedrene to the hydrocarbon follower. Chemistry of Natural Compounds. 2017;**53**:1-4. DOI: 10.1007/s10600-017-2085-1

[2] Nabeta K, Ara Y, Aoki Y, Miyake M. Biosynthesis of monoterpenes and sesquiterpenes in *Larix leptolepis* Callus from deuterated mevalonates. Journal of Natural Products. 1990;**53**:1241-1248. DOI: 10.1021/np50071a016

[3] Blagg BSJ, Jarstfer DH, Rogers DH, Poulter CD. Recombinant squalene synthase. A mechanism for the rearrangement of presqualene diphosphate to squalene. Journal of the American Chemical Society. 2002;**124**:8846-8853. DOI: 10.1021/ja020411a

[4] Paknikar SK, Kadam SH, Ehrlich AL, Bates RB. Alternate biosynthesis of valerenadiene and related sesquiterpenes. Natural Product Communications. 2013;**8**:1195-1196. PMID: 24273843

[5] Morales PA, Amaro-Luis JM, Kutateladze AG. Structure determination and mechanism of formation of a seco-moreliane derivative supported by computational analysis. Journal of Natural Products. 2017;**80**:1210-1214. DOI: 10.1021/acs.jnatprod.7b00041

[6] Joseph-Nathan P, Reyes-Trejo B, Morales-Rios MS. Molecular rearrangements of (−)-modhephene and (−)-isocomene to a (−)-triquinane. Journal of Organic Chemistry. 2006;**71**:4411-4417. DOI: 10.1021/jo060258t

[7] Uroos M, Lewis W, Blake AJ, Hayes CJ. Total synthesis of (+)-cymbodiacetal: A re-evaluation of the biomimetic route. Journal of Organic Chemistry. 2010;**75**:8465-8470. DOI: 10.1021/jo101691n

[8] Bottini AT, Dev V, Garfagnoli DJ, Hope H, Joshi P, Lohani H, Mathela CS, Nelson TE. Isolation and crystal structure of a novel dihemiacetal bis-monoterpenoid from *Cymbopogon martini*. Phytochemistry. 1987;**26**:2301-2302. DOI: org/10.1016/S0031-9422(00)84706-8

[9] D'Souza AM, Paknikar SK, Dev V, Beauchamp PS, Kamat SP. Biogenetic-type synthesis of (+)-Cymbodiacetal, a constituent of *Cymbopogon martini*. Journal of Natural Products. 2004;**67**:700-702. DOI: 10.1021/np030338h

[10] Heib AE, Benharref A, Parres-Manadie S, Manoury E, J-C Daran M, Urrutigoity M. Molecular rearrangement of epoxide derived from sesquiterpenes by Lewis acid catalysis. Tetrahedron: Asymmetry. 2010;**21**:1272-1277. DOI: org/10.1016/j.tetasy.2010.04.024

[11] Bisarya SC, Dev S. Studies in sesquiterpenes—XXXIII: Himachalol. Tetrahedron. 1968;**24**:3869-3879. DOI: org/10.1016/S0040-4020(01)92593-1

[12] Pai Fondekar KP, Malik BL, Paknikar SK. Santonic acid: Zn–HCl–ether reduction and ceric ammonium nitrate oxidation. Natural Product Research. 2014;**28**:1567-1573. DOI: org/10.1080/14786419.2014.927464

[13] Zhao B, Lin X, Lei L, Lamb DC, Kelly SL, Waterman MR, Cane DE. Biosynthesis of the sesquiterpene antibiotic albaflavenone in *Streptomyces coelicolor* A3(2). Journal of Biological Chemistry. 2008;**283**:8183-8189. DOI: 10.1074/jbc.M710421200

[14] Cane DE, Lin X. Biosynthesis of the sesquiterpene antibiotic albaflavenone in *Streptomyces coelicolor*. Mechanism and stereochemistry of the enzymatic formation of epi-isozizaene. Journal of the American Chemical Society. 2009;**131**:6332-6333. DOI: 10.1021/ja901313v

[15] Moody SC, Zhao B, Lei L, Nelson DR, Mullins JGL, Waterman MR, Kelly SL, Lamb D. Investigating conservation of the albaflavenone biosynthetic pathway and CYP170 bifunctionality in streptomycetes. FEBS Journal. 2012;**279**:1640-1649. DOI: 10.1111/j.1742-4658.2011.08447.x

[16] Kobayashi T, Kon Y, Abe H, Ito H. Concise total synthesis of albaflavenone utilizing sequential intramolecular aldol condensation: Determination of absolute configuration. Organic Letters. 2014;**16**:6397-6399. DOI: 10.1021/ol503202d

[17] Nguyen QNN, Tantillo DJ. Caryolene-forming carbocation rearrangements. Beilstein Journal of Organic Chemistry. 2013;**9**:323-331. DOI: 10.3762/bjoc.9.37

[18] Brock NL, Huss K, Tudzynski B, Dickschat JS. Genetic dissection of sesquiterpene biosynthesis by *Fusarium fujikuroi*. ChemBioChem. 2013;**14**:311-315. DOI: 10.1002/cbic.201200695

[19] Citron CA, Brock NL, Tudzynsky B, Dickschat JS. Labelling studies on the biosynthesis of terpenes in *Fusarium fujikuroi*. Chemical Communications. 2014;**50**:5224-5226. DOI: org/10.1039/C3CC45982A

[20] Khan VA, Gatilov YV, Dubovenko ZV, Pentegova VA. Crystal structure of koraiol—A sesquiterpene alcohol with a new type of carbon skeleton from the oleoresin of *Pinus*

koraiensis. Chemistry of Natural Compounds. 1979;**15**:572-576. DOI: org/10.1007/BF00565927

[21] Hinkley SFR, Perry NB, Weavers RT. Confirmation of structure and absolute stereochemistry of 9-epi-β-caryophyllene from *Dacrydium cupressinum*. Phytochemistry. 1994;**35**:1489-1494. DOI: org/10.1016/S0031-9422(00)86882-X

[22] Barra L, Schulz B, Dickschat JS. Pogostol biosynthesis by the endophytic fungus *Geniculosporium*. ChemBioChem Communications. 2014;**15**:2379-2383. DOI: 10.1002/cbic.201402298

[23] Amand S, Langenfeld A, Blond A, Dupont J, Ney B, Prado S. Guaiane sesquiterpenes from *Biscogniauxia nummularia* featuring potent antigerminative activity. Journal of Natural Products. 2012;**75**:798-801. DOI: 10.1021/np2009913

[24] Teisseire P, Maupetit P, Corbier B. Contribution to the knowledge of Patchouliu oil. Recherches (Paris). 1974;**19**:8-35

[25] Dolejs L, Mironov A, Sorm F. On terpenes. CXXI, structure of bulnesol and stereochemistry of guaiol, nepetalinic acids and iridomyrmecins. Collection of Czechoslovak Chemical Communications. 1961;**26**:1015-1020. DOI: org/10.1135/cccc19611015

[26] Van Beek TA, Joulain D. The essential oil of patchouli, *Pogostemon cablin*: A review. Flavour and Fragrance Journal. 2018;**33**:6-51. DOI: 10.1002/ffj.3418

[27] Crotea R, Munk SL, Akoh CC, Fisk HJ, Satterwhite DM. Biosynthesis of the sesquiterpene patchoulol from farnesyl pyrophosphate in leaf extracts of *Pogostemon cablin* (patchouli): Mechanistic considerations. Archives of Biochemistry and Biophysics. 1987;**256**:56-68. DOI: org/10.1016/0003-9861(87)90425-5

[28] Akhila A, Sharma PK, Thakur RS. Biosynthetic relationships of patchouli alcohol, seychellene and cycloseychellene in *Pogostemon cablin*. Phytochemistry. 1988;**27**:2105-2108. DOI: org/10.1016/0031-9422(88)80105-5

[29] Faraldos JA, Wu S, Chappell J, Coates RM. Doubly deuterium-labeled patchouli alcohol from cyclization of singly labeled [2-$^{2}H_{1}$]farnesyl diphosphate catalyzed by recombinant patchoulol synthase. Journal of the American Chemical Society. 2010;**132**:2998-3008. DOI: 10.1021/ja909251r

[30] Pyle BW, Tran HT, Pickel B, Haslam TM, Gao Z, MacNevin G, Vederas JC, Kim S-U, Ro D-K. Enzymatic synthesis of valerena-4,7(11)-diene by a unique sesquiterpene synthase from the valerian plant (*Valeriana officinalis*). FEBS Journal. 2012;**279**:3163-3146. DOI: 10.1111/j.1742-4658.2012.08692.x

[31] Yeo YS, Nybo SE, Chittiboyina AG, Weerasooriya AO, Wang YH, Gongoa-Castillo E, Vaillancourt B, Buell CR, DellaPenna D, Celiz MD, Jones AD, Wurtele ES, Ransom N, Dudareva N, Shaaban KA, Tibrewal N, Chandra S, Smillie T, Khan IA, Coates RM, Watt DS, Chappell J. Functional identification of valerena-1,10-diene synthase, a terpene synthase catalyzing a unique chemical cascade in the biosynthesis of biologically active

sesquiterpenes in *Valeriana officinalis*. Journal of Biological Chemistry. 2013;**288**:3163-3173. DOI: 10.1074/jbc.M112.415836

[32] Paul C, Koenig WA, Muhle H. Pacifigorgianes and tamariscene as constituents of *Frullania tamarisci* and *Valeriana officinalis*. Phytochemistry. 2001;**57**:307-313. DOI: org/10.1016/S0031-9422(01)00018-8

[33] Yan S, Li S, Wu W, Zhao F, Bao L, Ding R, Gao H, Wen HA, Song F, Liu HW. Terpenoid and phenolic metabolites from the fungus *Xylaria* sp. associated with termite nests. Chemistry and Biodiversity. 2011;**8**:1689-1700. DOI: 10.1002/cbdv.201100026

[34] Granchi C, Minutolo F. Anticancer agents that counteract tumor glycolysis. ChemMed Chem. 2012;**7**:1318-1350. DOI: 10.1002/cmdc.201200176

[35] Rabe P, Schmitz T, Dickschat JS. Mechanistic investigations on six bacterial terpene cyclases. Beilstin Journal of Organic Chemistry. 2016;**12**:1839-1850. DOI: 10.3762/bjoc.12.173

[36] Bhatawadekar SV, Pednekar PR, Chakravarti KK, Paknikar SK. A survey of sesquiterpenoids of vetiver oil. In: Atal CK, Kapur BM, editors. Cultivation and Utilization of Aromatic Plants. Jammu-Tawi: RRL, CSIR; 1982. pp. 412-426 and references cited therein

Chapter 6

Diterpenes from Different Fungal Sources and Their ^{13}C-NMR Data

Fozia Farhat, Arneeb Tariq, Annam Zikrea and
Riffat Nasim Fatima

Additional information is available at the end of the chapter

http://dx.doi.org/10.5772/intechopen.79186

Abstract

Diterpenes are one of the classes of natural products with about 7000 structures. The basic skeleton of diterpene contains 20 carbon atoms. Microbes contain a large number of diterpenoid with many oxidized carbons and nitrogen atoms. To date, a number of secondary metabolites have been isolated from fungal sources, and some of these examples showed diverse structural features and interesting biological activities. These classes of compounds have attracted the interest of natural product scientist due to their potential biological activities. This chapter includes recently (2013–2018) isolated compounds from various fungal sources especially cythane, clerodanes, halimanes, abietane, and indole-type diterpenes. Biosynthetic pathway of plants and fungi diterpenes showed homology at initial steps but showed differences at latter steps. The biological activity and ^{13}C-NMR data of these recently isolated compounds have been discussed. These diterpenes exhibited potential nitric oxide, anticancer, antioxidant, and antitumor properties. The diterpenes are clerodane, labdane, and kaurane derivatives. A brief discussion on the ^{13}C-NMR chemical shifts of these diterpenes has been discussed at the end of each type.

Keywords: fungal, biosynthesis, diterpenes, biological activities, ^{13}C-NMR

1. Introduction

Terpenoids comprise the largest, structurally most diverse family of natural products and play important roles in all living organisms [1, 2]. Fungi (*Ascomycota* and *Basidiomycota*) are prolific producers of structurally diverse terpenoid compounds. Classes of terpenoids identified in fungi include the sesqui-, di-, and triterpenoids.

As the largest group of documented natural products, terpenoids have attracted attention from a broad scientific community and have been heavily investigated due to their interesting structural characteristics and profound biological effects [3–6].

Fungi are important source of potential bioactive compounds which play an important role in pharmacology industry [7–11]. Among fungi, mushrooms are the most attractive sources of bioactive compounds both of chemical and biomedical interests. Approximately 2000 mushrooms are safe for human consumption, and about 650 of them have medicinal properties out of 15,000 documented species of mushrooms [12]. These are also important in industrial processes to enhance composition of bioactive compounds in fermented grain assays [13–15].

2. Diterpene biosynthesis

Diterpenoid biosynthesis has been studied in plants, bacteria, and fungi [16]; still lots of work are required to clone many important respective genes to characterize and engineer diterpenoid pathways in these representative organisms which remain a big challenge [17]. Fungal di-TPS enzymes show homology to plant enzymes in terms of its size and the combination of biochemical studies with molecular genetics. This also facilitated the comparison of plant and fungal biochemical pathways leading to the formation of gibberellins in plants and fungi [18].

The first committed step in diterpenoid biosynthesis is the cyclization of GGPP to produce the diterpene scaffold, which occurs via a carbocation cascade. Classically, activation of the carbocation cascade by terpene synthases corresponds to the removal of the pyrophosphate group from the linear substrate. This ionization-dependent reaction is catalyzed by class I terpene synthases [19]. Fusicoccanes are potent phytotoxins known to be synthesized by a few fungal species. *P. amygdali* was the first monofunctional diterpene synthase cloned and characterized in *E. coli* [20, 21]. Diterpenoids cyclized by the first, one-step route involves a monofunctional class I diterpene synthase that catalyzes ionization-dependent diphosphate cleavage and subsequent carbocation migration and quenching using a mechanism similar to sesquiterpene synthases, except the prenyl chain is now longer by one isoprene unit [22].

Biosynthesis of labdane-type diterpenoids requires a two-step cyclization pathway involving first a protonation dependent cyclization of GGPP to form the characteristic labdane bicycle and, in the second step, ionization-dependent cyclization at a separate active site to generate the final cyclic product (**Figure 1**). Cyclization of GGPP to ent-CDP and then to the tetracyclic ent-kaurene generates the precursor for gibberellin (gibberellic acids, GA) phytohormones that are major regulators of plant growth and development. It is believed that because of its essential role in plants, ent-kaurene represents the ancestral diterpenoid cyclization pathway from which alternative cyclization routes evolved to generate the large diversity of labdane-type compounds known today [22]. In fact, it has been shown that single amino acid changes are sufficient to alter the product profile of the class I ent-kaurene synthase to form new cyclic scaffolds [23, 24].

http://dx.doi.org/10.5772/intechopen.79186

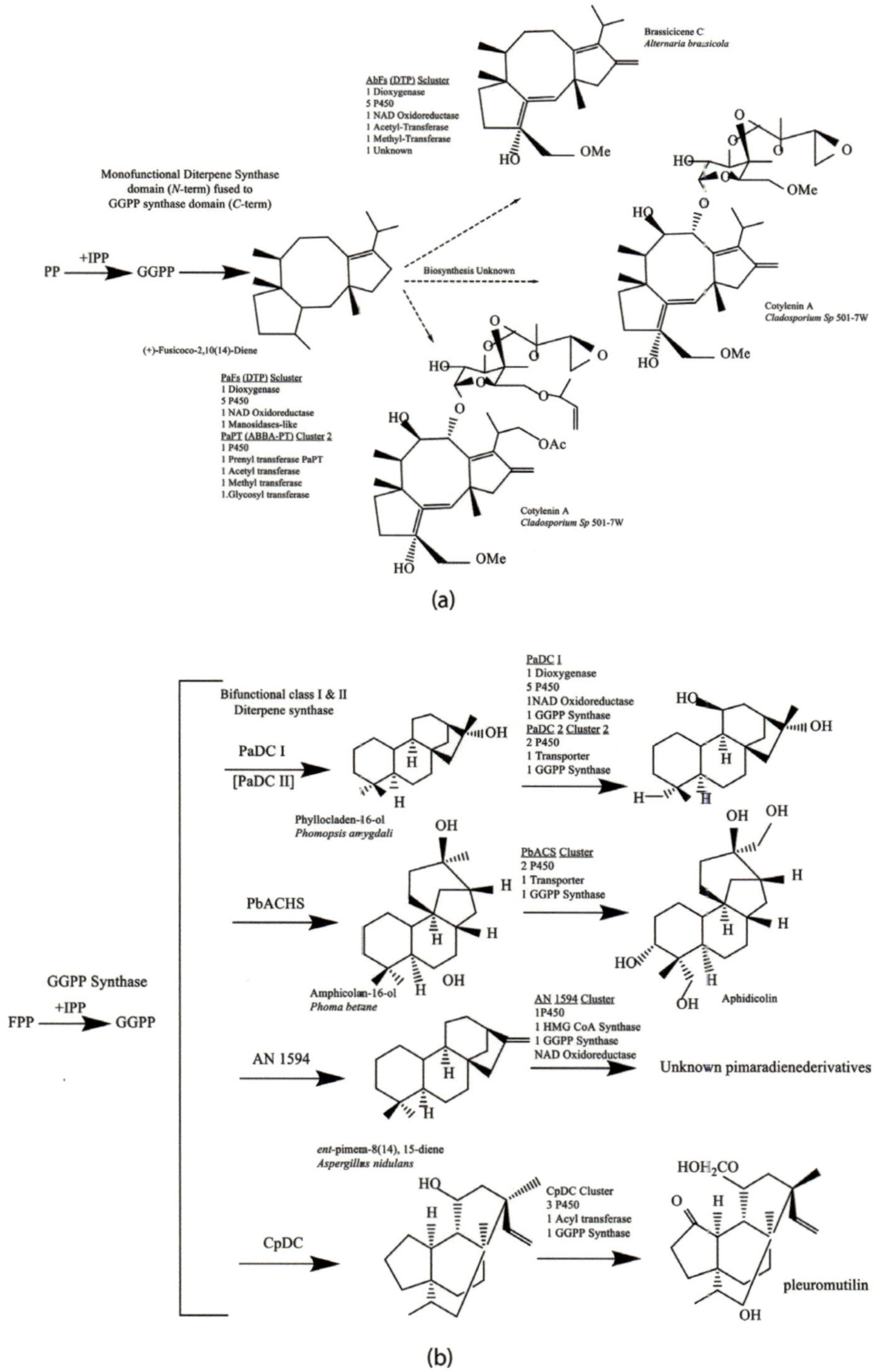

Figure 1. Overview of diterpenoid biosynthesis: (a) monofunctional-2,10(14)-diene is modified into different fusicoccane compounds and (b) bifunctional diterpene synthases make different labdane-related scaffolds that are modified into bioactive compounds.

3. Diterpenoid classification

3.1. Bicyclic diterpenoids

3.1.1. Clerodane diterpenes

Clerodane diterpenes are the natural group of secondary metabolites holding an utmost pharmacological significance. These are bicyclic structures consisting of a fused ring (decalin moiety from C-1 to C-10) with a six-carbon side chain (C-11 to C-16) attached at C-9. The rest of the carbons (C-17 to C-20) are bonded at C-8, C-4, C-5, and C-9, correspondingly [25]. Only 25% of the clerodane diterpenes showed 5:10 *cis* ring junction, while the rest possess 5:10 *trans* ring fusion as presented here in the form of columbin and clerodin, respectively. Columbin exhibited dose-dependent anti-inflammatory activity as well as chemopreventive activity against colorectal cancer [26–28]. During the last 25 years, over 1300 diterpenoids and nor-diterpenoids with the clerodane carbon skeleton have been isolated [29, 30]. The detailed classification of clerodane diterpenes is given in **Figure 2**.

3.1.1.1. Clerodane diterpenes by biotransformation from endophytic fungi

Three strains of endophytic fungi *L. gonubiensis*, *N. ribis*, and *P. stromaticum* produced one known and five unknown compounds (B1–B4) through a process of biotransformation, while compounds Q1–Q2 are derived as chemical derivatization of compound 2 [30]. These compounds were actually isolated for the first time from *Croton argyrophylloides* (Euphorbiaceae) and further biotransformed by *Cunninghamella echinulata* and *Rhizopus stolonifer* fungi and produced a new diterpene, as previously described by Monte et al. [31] and Mafezoli et al. [32] (**Figure 3**).

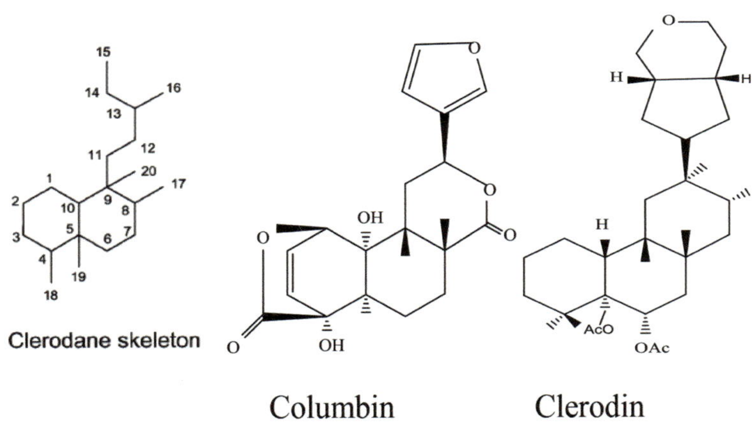

Figure 2. Clerodane skeleton, cis and trans structures of clerodane.

http://dx.doi.org/10.5772/intechopen.79186

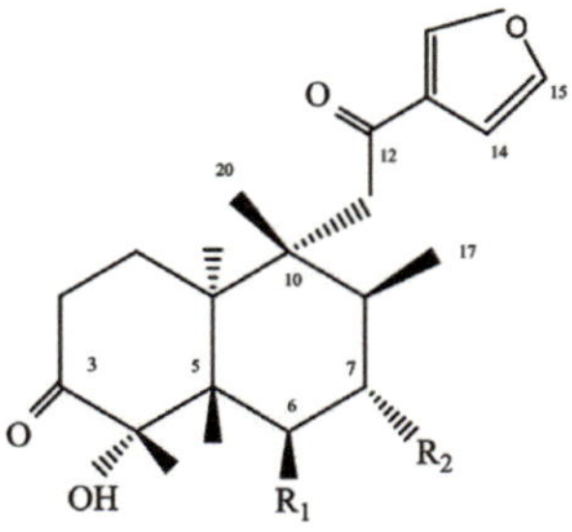

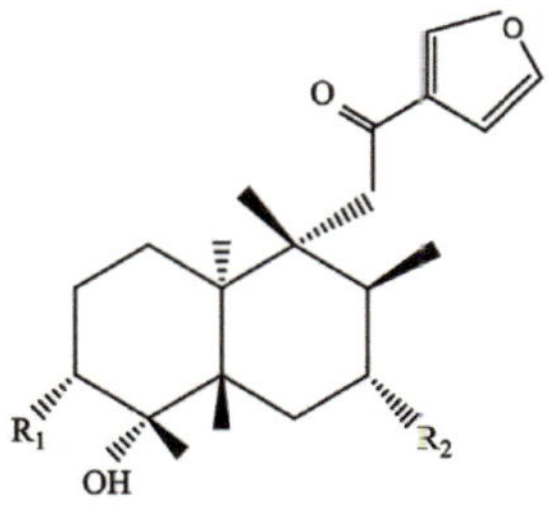

R_1=H, R_2=H (Substrate)

R_1=H R_2=OH (Compound B2)

R_1=H R_2=$OCOCH_2CH_3$ (compound Q1)

R_1=H R_2=OCH_2Ph (compound Q2)

R_1=OH R_2=H (compound B4)

1. R_1=OH, R_2=H (B1)

R_1=OH R_2=OH (compound B3)

[31]

Figure 3. Chemical structures of the metabolites by and chemical derivatives of the (3*R*,4*S*,5*S*,8*S*,9*R*,10*S*)-3,12-dioxo-15,16-epoxy-4-hydroxycleroda-13(16),14-diene (compound 1) [35].

3.1.1.2. 13C-NMR date

13C-NMR spectra of substrate 1 and B2 suggested the C-7 hydroxylation making the signal at δ 70.1 (CH) in B2. Compound B2 is identified as new metabolite (4S,5S,7R,8R,9S,10S)-4,7-dihydroxy-15,16-epoxy-3,12-dioxocleroda-13(16),14-diene, and its molecular formula $C_{20}H_{28}O_5$ is sorted by HRMS. Compound B3 was unique for cultures of *N. ribis*. The 13C-NMR spectrum of B3 showed the presence of one at δ 72.2 in the spectrum confirmed that compound 1 was regioselectively bioreduced at C-3. The new compound B3 was named (3*R*,4*S*,5*S*,7*R*,8*S*,9*R*,10*S*)-3,4,7-trihydroxy-15,16-epoxy-12-oxocleroda-13(16),14-diene, which is in agreement with the molecular formula $C_{20}H_{30}O_5$. The biotransformation product B4 was obtained only in the *P. stromaticum* culture. The 13C-NMR spectrum of B4 showed no reduction of carbonyl group at δ 213.5 (C-3) and the appearance of carbinol methane group at δ 71.8 (C-6). And it is named as (4S,5R,6R,8S,9R,10S)-4,6-dihydroxy-15,16-epoxy-3,12-dioxocleroda-13(16),14-diene, which is in agreement with the molecular formula $C_{20}H_{28}O_5$. The new compound Q1 was named (4S,5S,7R,8R,9S,10S)-7-propionyloxy-4-hydroxy-15,16-epoxy-3,12-dioxocleroda-13(16),14-diene. The new derivative Q2 was named (4*S*,5*S*,7*R*,8*R*,9*S*,10*S*)-7-benzyloxy-4-hydroxy-15,16-epoxy-3,12-dioxocleroda-13(16),14-diene [30].

3.1.1.3. Biological activity

Clerodane diterpenes possessed effective insect antifeeding and related insecticidal properties. There are approximately more than 400 natural and semisynthetic products that have been assayed in the laboratories showing potential antifeedant properties [33].

3.1.2. Labdanes

The labdane-related diterpenoids are a special group, consisting of over 7000 members, which are distinguished by their unique biosynthesis. Gibberellin phytohormones as well as antibiotics such as some of phytoalexins and phytoanticipins fall into this family [34]. Labdanolic acids have been identified as biomarkers for the botanical origin of French ambers [35], while copalic acid and its relatives have been associated with the biological activity of the resins from *Copaifera* species. The lanceolatins are a group of **labdanes** and **abietanes** which were obtained [40] from *Cephalotaxus lanceolata* (Cephalotaxaceae). Some of the abietanes were described in this chapter as well (**Figure 4**).

3.2. Tricyclic diterpenoids

See **Figure 5**

3.2.1. Abietanes

Abietane is a class of diterpenoids with excellent metabolic profile. Compounds of this class showed broad spectrum antiviral, antibacterial, and antifungal activity [36, 37]. Abietane diterpenoids are extracted from few fungal species [38]. The intervention of quinone methides in the antioxidant activity of the phenolic diterpenoids ferruginol and carnosic acid has been discussed. The antifungal activity of some abietic acid esters in the context of their use as wood preservatives and the antiviral activity of podocarpic acid derivatives have been examined. In case of human cells, antiproliferative effect on tumor cells has been reported [39].

About 200 compounds of this family have been identified commonly known as dehydroabietic derivatives (dehydroabietanes) [40] assuming 20-carbon saturated aromatic ring I, abietane as standard (**Figure 6**).

3.2.1.1. Tricyclic abietatrienes

This group of abietane terpenes includes a tricyclic ring, three double bonds on B or C rings. Carboxylic acids are representatives of this group, of which the earliest example is the biologically active dehydroabietic acid (**Figure 6**), which possess an acid group at C-18 [42].

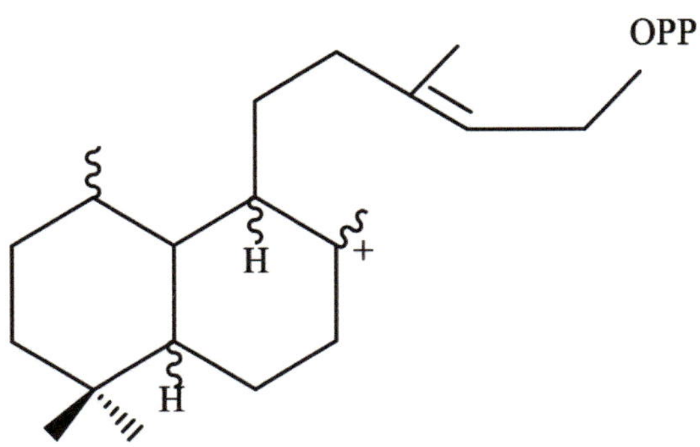

Figure 4. Skeleton of labdanes.

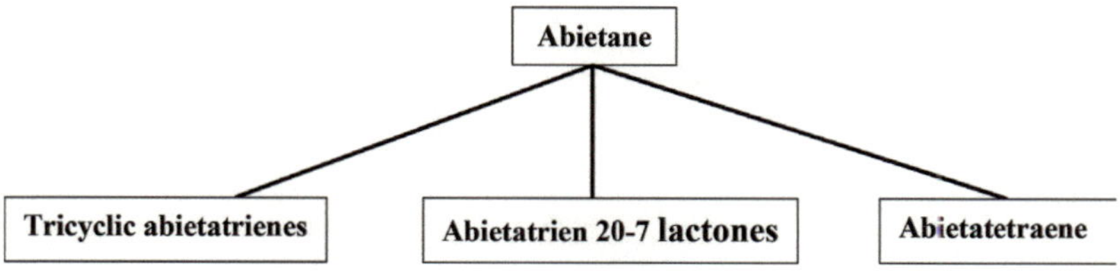

Figure 5. Classification of tricyclic abietane diterpenes.

3.2.1.1.1. Abietic acid

Abietic acid was extensively studied in various organisms for its biological activities. Current data also suggest that abietic acid is an important compound for synthesis of novel metabolites. They contribute to the body of knowledge related to compound 1 and deepen the understanding of the potential and properties of 1 and its derivatives (**Figure 7**).

3.2.1.1.2. Dehydroabietic acid

Dehydroabietic acid displays not only antiulcer and antimicrobial properties but also antitumor and anti-inflammatory effects (compound 2). Antimicrobial effects of DHA have been studied, specifically against methicillin-resistant strains of *Staphylococcus aureus* [43].

Biological activity: It also showed activity against other Gram-positive organisms such as *Salmonella* sp., *Bacillus subtilis*, and *E. coli* [44]. This latter study also described the inhibition of nitric oxide (NO) production by DHA, which was reported by other researchers as well [49]. Kawada et al. [45] have reported in relation with the inhibition of pro-inflammatory cytokines that DHA is useful for treating obesity-related diseases.

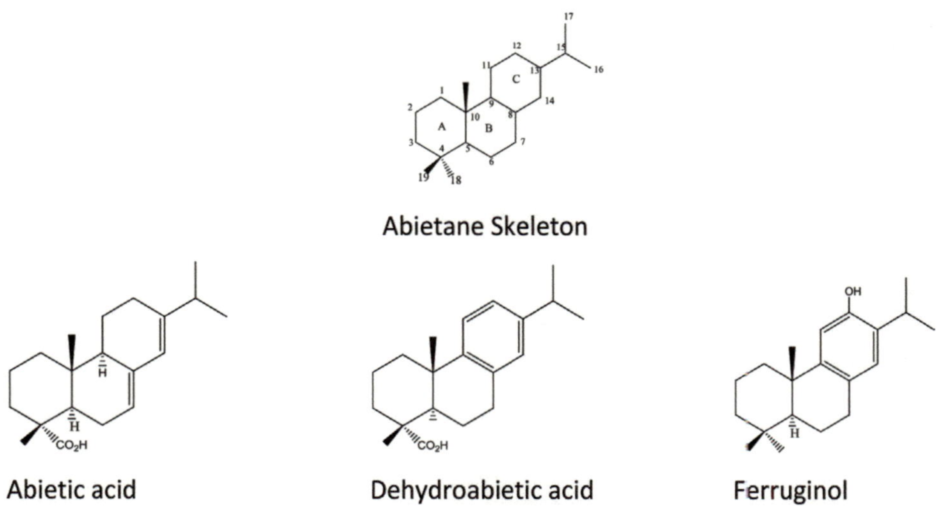

Figure 6. Abietane skeleton with some standard compounds [41].

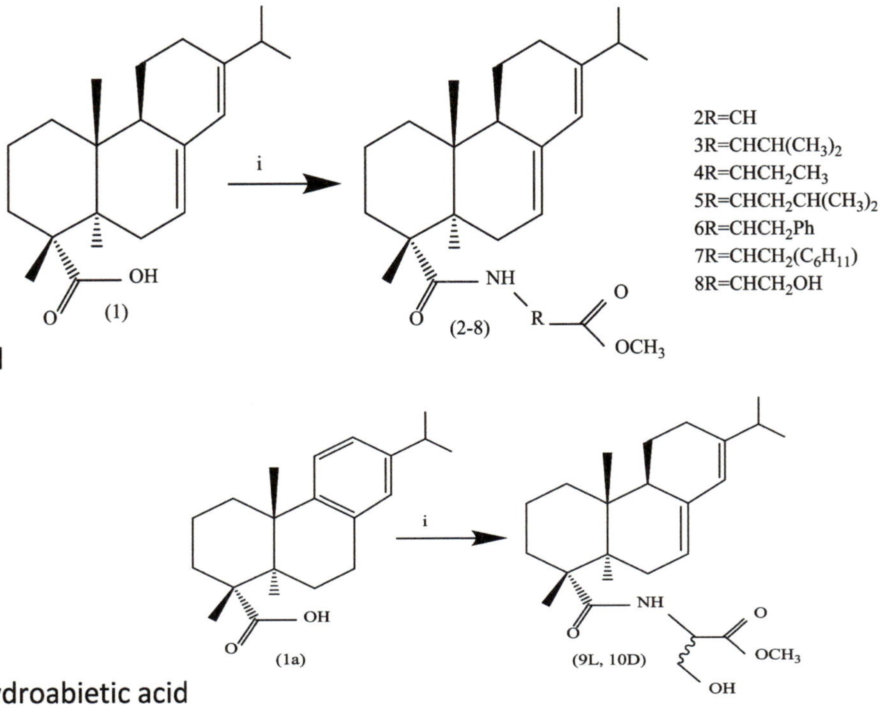

Abietic acid

Dehydroabietic acid

Figure 7. Synthesis of compounds 2–10 [42].

3.2.1.1.3. Ferruginol

Ferruginol (abieta-8, 11, 13-triene-12-ol) is the simplest phenolic abietane diterpenoid (3). This abietane occurs in plants belonging to the Podocarpaceae and Lamiaceae families [22].

Biological activity: This diterpene has attracted much attention since it has exhibited important bioactivities, such as antimicrobial [46], miticidal [47], cardioactive [48], and antioxidative [49]. Moreover, it accelerates the gastric ulcer healing process, and such effects have been related with the ability of ferruginol to increase the gastric prostaglandin content in vitro [50–52].

3.2.1.1.4. Callitrisic acid

Callitrisic acid is a diterpenoid acid contained in the resins of several *Callitris* species (Cupressaceae). It was simultaneously reported as a new natural product [53, 54]. This acid also occurs in plants of the genus *Juniperus* and *Calceolaria*, and it has also been found in the genus *Illicium*. Recently, a series of related acids to callitrisic acid having a C-19 carboxylic group have been isolated [55].

Biological activity: All these acids demonstrated important antiviral activity and significant anti-inflammatory activity [56].

3.2.2. Abietatriene 20-7 lactones

The abietatriene lactones are a group of compounds which possess an oxygen-containing ring which predominantly is in the form of lactones (i.e., abietatrien-20,7-olides). This group of abietanes are exemplified by carnosol (11,12-dihydroxy-8,11,13-20,7-olide). Carnosol possesses an aromatic C ring, carbon C-20 is a keto group, and carbons C-11 and C-12 are hydroxy groups. This abietane has displayed several biological activities. It displayed antioxidant, antimicrobial, anti-inflammatory, antitumor, and anti-HIV (IC50 = 8.0 μM) properties [57, 58].

3.2.3. Abietatetraenes

The abietatetraenes are a group of compounds which possess a fourth double bond which can be located at different positions. Among the 5,6-dehydro derivatives are coleon C and coleon U and related compounds (**Figure 8**). These metabolites are common in plants of the genus *Coleus* (synonym *Plectranthus*) and have described to possess antitumor, antimicrobial, and antiproliferative activity [59, 60] (**Table 1**).

3.3. Tetracyclic diterpenoids

3.3.1. Cythane diterpenes

Cyathus is a genus of fungi in the Nidulariaceae, a family collectively known as the bird's nest fungi. Such compounds are named so, as they resemble tiny bird's nests filled with "eggs," structures large enough to have been mistaken in the past for seeds. The first cyathin A_3 and allocyathin B_3 were reported from fungus *C. helenae* in 1972, and since then a number of other diterpenes being isolated and documented from different species belong to genus *Cyathus* [61]. In particular, the species belonging to the genus *Cyathus* is recognized as prolific producer of bioactive cyathane diterpenoids with inimitable tricyclic ring skeleton [62]. Cyathane diterpenoids also represent a group of natural products with versatility both in structure and bioactivity [63, 64]. Cyathane diterpenes are important bioactive metabolites extracted from the genus *Cyathus, Hericium,* and *Sarcodon.* Genus *Cyathus* is beneficial in producing healthy food and possesses the potential of nitric oxide (NO) inhibition and antibacterial activities [64].

3.3.1.1. Diversity of cythane diterpenes

A number of other biologically significant cyathane diterpenoids have been isolated from the fruiting bodies of mushroom *Sarcodon scabrosus* [65–67], *Sarcodon glaucopus* [68, 69], and *Sarcodon cyrneus* [70, 71] and the culture of fungi *C. helenae* [72], *C. africanus* [73], *C. earlei*

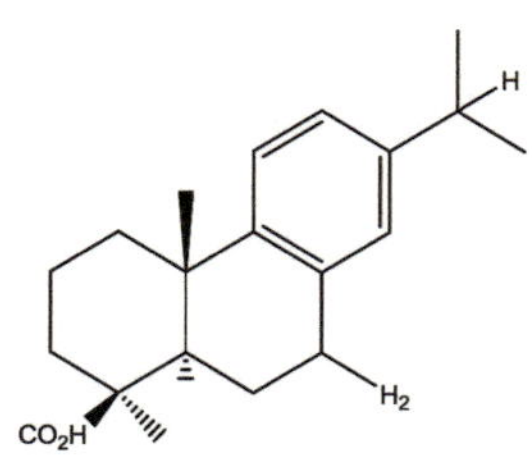

Callistric acid (11)

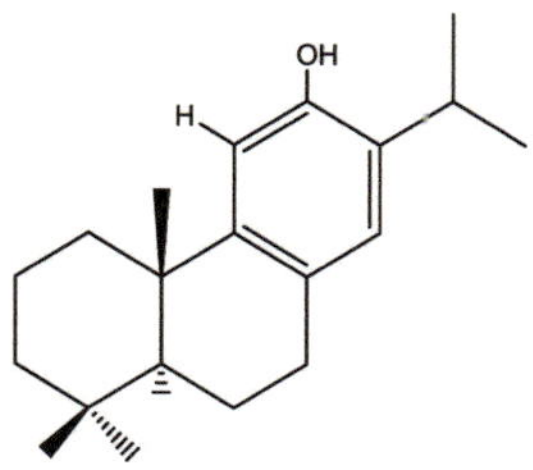

Pisiferic acid (12)

Figure 8. Representative member of tricyclic abietatrienes *Callitris* species.

Carbon number	Compound B2[a]	Compound B3[b]	B4[b]	Q1[b]	Q2[b]	2[b]	3[b]	4[b]	5[b]	6[a]	7[b]	8[b]	9[b]	10[b]
1	24.6	21.4	23.5	23.5	23.5	14.3	14.2	9.8	14.3	14.3	14.3	14.3	16.3	16.3
2	37.6	30.2	35.8	36.1	36.1	17.1	17.1	14.3	17.1	17	17.1	17.1	18.6	18.6
3	215.7	72.2	213.5	214.4	214.3	18.4	18.1	17.1	18.4	18.4	18.4	18.4	21.1	21.0
4	83.0	79.0	83.0	81.4	81.3	21.0	18.4	18.4	21.0	21.0	21.0	21.0	23.9	23.9
5	46.7	43.1	48.9	45.6	45.6	21.6	19.1	21.0	21.6	21.6	21.6	21.6	25.2	23.9
6	41.7	41.4	71.9	37.3	37.1	22.7	20.9	21.6	22.2	22.6	22.6	22.6	30.0	25.2
7	70.1	70.0	34.4	72.2	77.4	25.4	21.5	22.6	22.6	25.3	25.4	25.5	33.4	30.0
8	45.9	44.9	35.3	41.8	41.8	27.6	22.6	25.4	23.0	27.6	26.3	27.6	37.1	33.4
9	44.0	42.8	42.0	42.7	42.7	34.8	25.5	25.6	25.2	34.8	26.4	34.8	37.2	37.1
10	42.7	42.0	41.2	41.0	40.9	35.0	27.5	27.6	25.4	35.0	26.5	35.1	37.9	37.2
11	48.3	47.4	46.8	47.1	47.0	37.7	31.3	34.8	27.6	37.7	27.6	37.5	45.6	37.9
12	197.4	194.8	196.8	194.5	194.3	38.5	34.7	35.0	34.8	38.1	32.7	38.4	47.4	45.6
13	131.0	129.6	129.6	129.5	129.5	41.7	35.0	37.7	35.0	38.4	33.7	45.9	52.6	47.4
14	109.4	108.8	108.8	108.8	108.8	45.9	37.8	38.4	37.8	45.7	34.6	46.6	54.8	52.6
15	146.1	144.5	144.7	144..6	144.7	46.5	38.4	45.8	38.5	46.5	34.8	51.1	63.6	54.8
16	149.9	147.0	147.1	147.0	147.1	51.1	45.6	46.5	41.8	51.1	35.0	53.0	123.9	63.5
17	12.2	11.8	16.3	11.6	11.6	52.4	46.6	51.1	45.8	52.4	37.7	55.3	124.0	123.9
18	22.2	16.0	22.0	22.0	21.9	120.7	51.0	52.14	46.5	53.4	38.5	63.9	126.8	124.0
19	16.7	16.1	9.4	15.9	19.0	122.6	52.1	53.6	51.1	120.7	40.1	120..5	134.5	126.8
20	19.2	18.8	17.8	19.0	15.8	135.7	57.3	120.6	52.3	122.6	45.8	122.5	145.7	134.5
21						145.3	120.5	122.6	120.7	127.3	46.5	135.7	146.7	145.7
22						170.3	122.6	135.7	122.6	128.8	50.4	145.5	171.1	146.8
23						178.8	135.6	145.3	135.7	129.4	51.1	171.3		171.2

http://dx.doi.org/10.5772/intechopen.79186

Carbon number	Compound B2[a]	Compound B3[b]	B4[b]	Q1[b]	Q2[b]	2[b]	3[b]	4[b]	5[b]	6[a]	7[b]	8[b]	9[b]	10[b]
24							145.2	173.4	145.3	135.6	52.4	179.4		179.1
25							172.9	178.2	174.0	136.3	120.6			
26							178.2		178.3	145.2	122.6			
27										172.6	135.7			
28										178.1	145.3			
29											174.1			
30											178.2			
Ref	[30]	[30]	[30]	[30]	[30]	[42]	[42]	[42]	[42]	[42]	[42]	[42]	[42]	[42]

[a] ^{13}C-NMR at 125 MHz.
[b] ^{13}C-NMR at 75 MHz in $CDCl_3$.

Table 1. ^{13}C-NMR data of New Clerodane Diterpenes from Fungal Biotransformation of the3,12-Dioxo-15,16-Epoxy-4-Hydroxycleroda-13(16),14-Diene and abietane diterpenoids.

[74], *C. striatius* [75], *Strobilurus tenacellus* [76], and *Hericium erinaceus* [77–84]. Some cyathane diterpenoids represented interesting and significant biological activities.

3.3.1.1.1. Cythane diterpenes from Cyathus gansuensis

Cyathus gansuensis was reported in 2002 and produced valuable bioactive metabolites from fermented grains of barley and rice [13] by transformation. Recently, seven new [85] metabolites (8–14) named have been isolated from fruiting body of *C. gansuensis* as presented in **Figure 9**. The L69 fungal strain was used to isolate these compounds (8–14).

Biological activity: NO inhibition activity was tested on mouse monocyte, macrophages. Seven newly discovered cyathane diterpene derivatives showed inhibitory activity against the NO production in LPS-activated macrophages. The fungus can be a good choice for a transformation on a large scale to acquire enough pure metabolites for the future [85].

^{13}C-NMR structural elucidation: The detail of ^{13}C-NMR is presented in **Table 2**. ^{13}C-NMR data for compounds 8–14 revealed 20 carbons ascribable for 4 methyls, 4 methylenes (one oxygenated), 4 methines (two oxygenated), two quaternary carbons, and six sp^2 carbons. According to NMR and HRTOFMS at m/z 341.2079, [M + Na] + presented molecular formula of 8 and 9 (cyathin J and K) as $C_{20}H_{30}O_3$ (six degrees of unsaturation), 10 (cyathin L) $C_{22}H_{32}O_5$ (seven degrees of unsaturation degrees), 11 (cyathin M) $C_{20}H_{30}O_5$ (six degrees of unsaturation), 12 (cyathin N) $C_{20}H_{28}O_5$ (seven degrees of unsaturation), 13 (cyathin O) $C_{20}H_{30}O_5$ (six degrees of unsaturation), and 14 (cyathin P) $C_{20}H_{28}O_5$ (seven degrees of unsaturation) [85].

3.3.1.1.2. Cythane diterpenes from Cyathus africanus

Cyathus africanus is a medicinal basidiomycete fungus. Diterpenes have been reported to possess multiple bioactivities consisting of antimicrobial and anti-inflammatory properties [86]. The presently reported metabolites in this text are collected by the study of various scientists. Moreover, they have been characterized on the basis of their structural elucidation by spectroscopic methods and discussed in detail in this chapter (**Table 2**). Some of the new metabolites documented by various scientists are isolated from *C. africanus*. Cyathin Q (15) an important metabolite (**Figure 10**) showed autophagy-dependent apoptosis [87]. The gene sequence of this

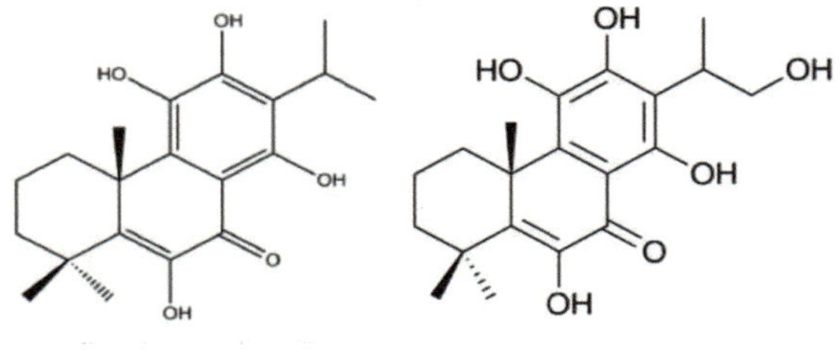

Coleon U (13) **Coleon C (14)**

Figure 9. Representative members of abietatetraenes.

Carbon number	Compound 8[a]	9[a]	10[a]	11[a]	12[a]	13[a]	14[a]	15[c]	16[c]	17[c]	18[c]	19[c]	20[c]	21[c]	22[c]	23[c]
1	145.9	145.9	39.8	83.7	84.3	53.4	84.5	38.6	39.9	38.8	34.9	33.8	33.7	83.8	82.1	53.0
2	129.0	129.2	29.4	209.7	209.0	211.1	209.6	29.3	29.7	29.7	23.7	23.2	22.8	209.8	75.3	210.6
3	142.3	142.9	140.4	141.8	140.7	145.0	142.6	140.8	144.9	141.8	79.2	77.4	77.5	141.9	140.7	144.4
4	146.2	147.3	139.2	176.2	177.0	177.1	172.5	140.6	144.0	139.9	78.5	77.0	76.6	176.4	146.7	179.6
5	44.1	41.6	41.7	46.8	142.6	41.0	41.4	38.5	152.7	36.5	36.0	37.6	31.8	47.0	45.8	45.6
6	45.7	44.8	45.1	45.0	49.4	46.5	45.3	47.2	48.6	45.1	46.2	43.6	42.8	45.2	43.4	58.2
7	29.2	34.1	31.0	28.9	29.7	29.5	29.8	35.8	33.8	35.1	33.7	28.0	32.3	29.1	28.9	35.5
8	33.6	34.1	38.4	37.2	35.3	37.3	35.2	36.2	38.0	17.5	31.2	31.7	30.8	37.4	38.1	39.0
9	55.8	55.8	50.6	47.7	47.6	43.0	47.1	51.2	50.3	50.9	43.6	42.6	42.2	47.9	50.9	43.6
10	29.1	38.5	30.5	28.2	124.4	27.6	32.2	30.3	121.3	32.1	24.8	26.2	25.7	28.4	29.2	27.0
11	130.0	72.6	72.0	128.6	118.2	74.6	76.6	160.2	135.8	72.2	159.9	160.5	69.9	128.8	129.8	39.2
12	143.1	147.8	129.7	143.7	149.6	55.9	49.3	144.7	130.1	146.8	145.2	142.4	146.0	143.8	143.3	156.5
13	69.2	125.9	150.5	69.1	71.4	71.8	70.2	78.5	33.5	159.1	78.2	79.5	157.6	69.3	69.1	127.3
14	81.5	76.7	76.2	80.6	81.9	106.4	107.4	75.5	76.5	77.4	74.8	80.2	76.0	80.8	91.3	209.2
15	66.6	64.8	169.4	66.3	63.1	60.6	62.2	196.2	172.1	194.7	196.1	194.2	192.9	66.5	66.6	26.9
16	17.8	17.3	17.2	17.8	24.9	13.6	13.4	17.7	27.0	16.7	18.9	19.7	17.7	18.0	18.0	16.4
17	19.2	19.3	25.1	20.3	21.0	26.3	22.3	24.8	24.2	24.7	19.9	20.0	19.5	20.4	20.6	23.9
18	27.4	27.4	28.4	27.0	26.8	26.3	26.3	28.5	28.2	28.5	29.1	28.1	28.3	27.2	28.4	27.2
19	22.6	23.2	21.9	19.8	20.5	20.7	20.2	21.8	21.9	22.2	19.7	19.4	19.3	20.0	22.5	19.8
20	23.4	23.4	22.4	20.3	21.3	20.8	21.4	22.6	22.0	22.6	19.9	19.6	19.8	21.6	23.2	21.2
21			172.3					28.0		57.3	57.9	59.3	57.1			
22			21.1													
Ref.	[85]	[85]	[85]	[85]	[85]	[85]	[85]	[87]	[88]	[88]	[88]	[88]	[88]	[89]	[89]	[89]

Carbon number	Compound 8[a]	9[a]	10[a]	11[a]	12[a]	13[a]	14[a]	15[c]	16[c]	17[c]	18[c]	19[c]	20[c]	21[c]	22[c]	23[c]
Carbon number	**Compound 24e**	**25[d]**			**26[d]**	**27[d]**	**28[d]**	**29[d]**	**30[d]**		**31[d]**	**32[d]**	**33[d]**	**34[f]**	**33[a]**	
1	213.7	53.1			88.5	47.0	82.8	84.5	90.5		215.4	216.0	53.3	38.5	38.3	
2	127.0	211.1			83.5	72.8	37.2	209.5	83.3		125.8	126.4	210.9	28.4	29.1	
3	188.5	144.2			140.2	77.8	137.2	141.6	139.6		192.3	193.2	144.9	139.9	140.7	
4	84.0	181.7			141.4	78.2	137.8	174.0	140.1		53.3	53.5	177.0	136.6	139.8	
5	43.7	43.2			39.0	37.1	41.6	43.3	41.3		38.9	37.5	40.5	40.4	36.1	
6	52.9	45.8			56.6	54.6	42.6	44.9	42.8		42.7	41.9	46.0	40.6	44.5	
7	33.1	32.0			34.8	34.7	31.0	30.5	30.9		29.7	29.9	29.7	30.4	34.3	
8	34.5	39.3			30.2	33.7	36.2	35.3	36.1		30.3	30.8	37.1	37.0	37.1	
9	54.0	43.5			48.7	42.6	49.9	47.3	47.2		50.2	50.4	42.9	49.2	50.4	
10	34.8	36.6			37.0	32.3	28.3	27.6	28.2		29.9	33.3	31.5	30.1	35.0	
11	73.0	71.6			72.4	72.3	80.0	79.9	79.9		74.6	76.9	76.3	72.4	62.1	
12	157.5	145.9			157.0	157.3	149.0	149.2	149.2		64.0	49.7	49.8	138.6	148.2	
13	122.6	127.3			123.2	123.2	126.6	126.9	126.6		57.0	70.6	70.2	158.2	154.6	
14	209.7	76.2			210.6	210.1	111.2	110.8	111.1		104.8	108.2	107.4	85.4	85.3	
15	64.4	65.0			64.5	64.4	58.9	58.9	58.9		59.1	62.1	62.0	192.9	194.2	
16	17.4	17.1			15.6	17.9	12.0	12.2	12.2		14.6	16.1	13.2	16.4	16.4	
17	14.5	24.3			23.6	20.7	17.3	22.4	19.6		20.5	22.4	26.0	24.5	24.7	
18	31.2	26.7			28.2	29.8	27.5	26.3	27.2		33.0	33.0	26.2	27.0	27.7	
19	22.2	21.0			24.3	20.5	21.2	20.2	19.3		23.4	21.0	20.5	21.5	22.0	
20	25.3	20.1			19.6	20.4	22.6	21.2	24.6		21.0	23.3	20.6	21.8	22.3	
21														56.6	***	
1′														105.3	106.3	

http://dx.doi.org/10.5772/intechopen.79186

Carbon number	Compound 8[a]	9[a]	10[a]	11[a]	12[a]	13[a]	14[a]	15[c]	16[c]	17[c]	18[c]	19[c]	20[c]	21[c]	22[c]	23[c]
2′														73.5	74.6	
3′														69.6	70.3	
4′														75.5	73.3	
5′														65.1	65.0	
Ref.	[90]	[90]			[90]	[90]	[90]	[90]	[90]		[90]	[90]	[90]	[96]	[96]	

[a] ^{13}C NMR at 125 MHz.
[b] ^{13}C NMR at 75 MHz in CDCl3.
[c] ^{13}C NMR at 150 MHz CD3OD.
[d] ^{13}C-NMR spectroscopic data for compounds 24–33 in MeOH at 200 MHz.
[e] ^{13}C-NMR at 125 MHz.
[f] ^{13}C-NMR 175 MHz in acetone-d6.

Table 2. ^{13}C-NMR data of cyathane diterpenoids.

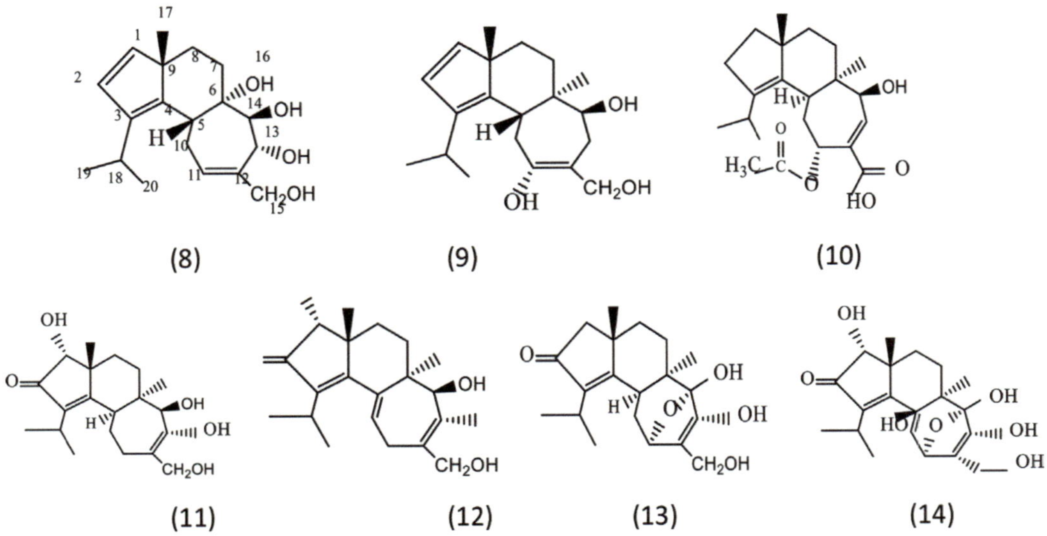

Figure 10. Newly (8–14) isolated metabolites from *C. gansuensis* [85].

strain has also been reported and submitted to GenBank with an accession numbers JX103204. Sequences of analysis exhibited 100% homology with that of fungus *C. africanus*. Compounds 16–20 (D–H) structurally represented new group of metabolites, while neosarcodonin O (21), cyathatriol (22), and 11-O-acetylcyathatriol (23) are also known cyathane diterpenes. Five novel compounds are isolated from *C. africanus* and show potential NO inhibition and cytotoxicity against HeLa cell line in vitro [88]. The structural elucidation is also described in **Figure 11**.

Ten new polyoxygenated cyathane diterpenoids, named as neocyathins (24–33), together with four known diterpenes are isolated from fungus *Cyathus africanus* (**Figure 12**). These compounds were isolated and identified by ^{13}C-NMR technique [90] (**Figure 13**).

Biological activity: Diterpenes with diverse bioactivities have been identified from plants and fungi [91]. Cyathin Q has the capacity to induce the apoptosis in HCT116 cells in a time- and dose-dependent manner. It was observed, when HCT116 cells exposed to 10 mM cyathin Q for 24 h exhibited apoptotic cells 82.07% [87]. This compound induced hallmarks of apoptotic events in HCT116 cells, including caspase activation, cytochrome c release, poly (ADP-ribose) polymerase (PARP) cleavage, and depolarization of the mitochondrial inner transmembrane potential. Nitric oxide has the capacity to react with aqueous oxygen,

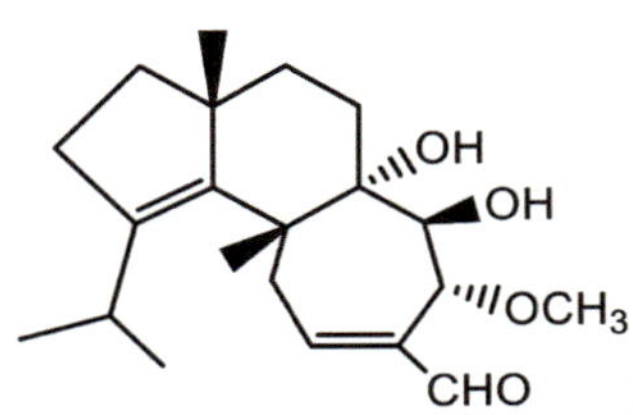
(15)

Figure 11. Structures of cyathane Q (15) isolated from *C. africanus* [87].

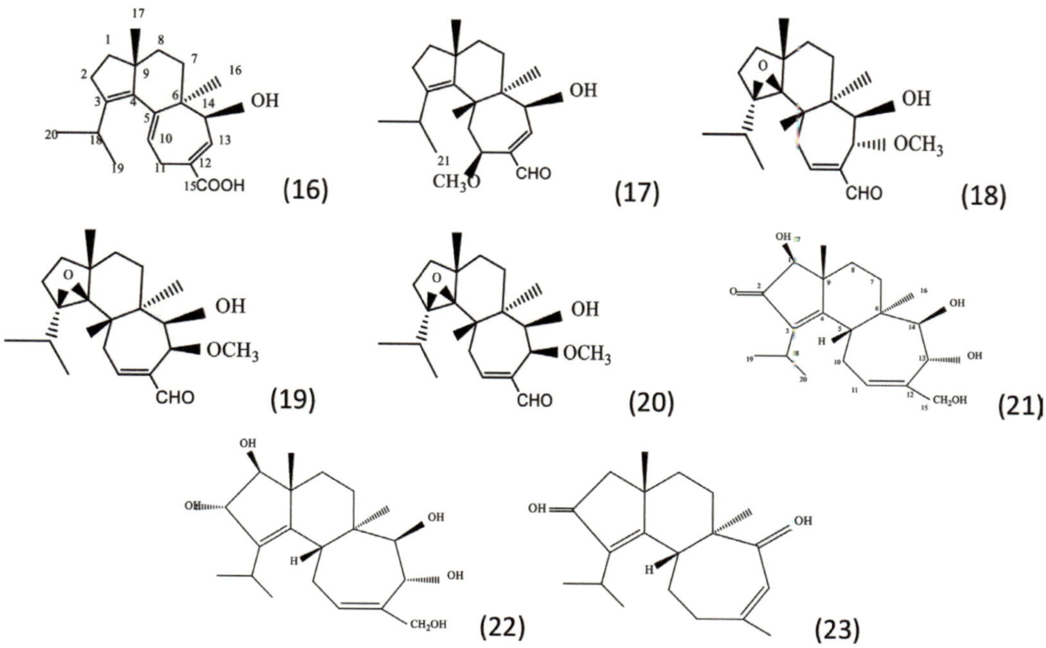

(16) (17) (18) (19) (20) (21) (22) (23)

Figure 12. Structures of metabolites isolated from *C. africanus* [88, 89].

superoxide, and transition metals like iron or zinc-sulfur clusters, and overproduction of NO is involved in many pathogenic diseases, including inflammation and cancer. The inhibition of NO overproduction in cells may prevent the occurrence of inflammatory diseases and cancer. The inhibition capacity (IC50) was more pronounced for 16, 17, and 19 by exhibiting NO inhibition 79.44, 89.2 and 84.33% reduction, respectively [91]. Moreover, inhibition of NO is concentration dependent as compounds 16–23 showed no NO inhibition at concentration 100 μM [88, 89]. COX-2 and iNOS are two major inflammatory mediators in brain neurodegeneration [92, 93]. Compounds isolated from *C. africanus* [90] showed strong COX-2 and iNOS capacities. Western blot analysis demonstrated that compounds 24 and 28 significantly suppressed LPS-induced COX-2 expression, whereas compounds 27, 28, 30, 31, and 33 markedly inhibited LPS-induced iNOS expression. Among these compounds, 28 showed strong inhibitory effects on both COX-2 and iNOS. Interestingly, 30 abolished LPS-induced iNOS expression but did not affect LPS-induced COX-2 expression. In addition, we also assayed the activities of iNOS enzyme [90].

^{13}C-NMR structural elucidation: The ^{13}C-NMR spectrum of some of the compounds isolated from *C. africanus* presented in **Table 2**.

3.3.1.1.3. Cythane diterpene from Hericium erinaceus and H. flagellum

Hericium genus is among the most blessed medicinal and eatable mushrooms and known to produce secondary metabolites with the potential to treat neurodegenerative diseases. It enables improvement of many brain-related disorders [94]. In this regard, neurotrophins are nerve growth factor (NGF) and brain-derived neurotrophic factor (BDNF) involved

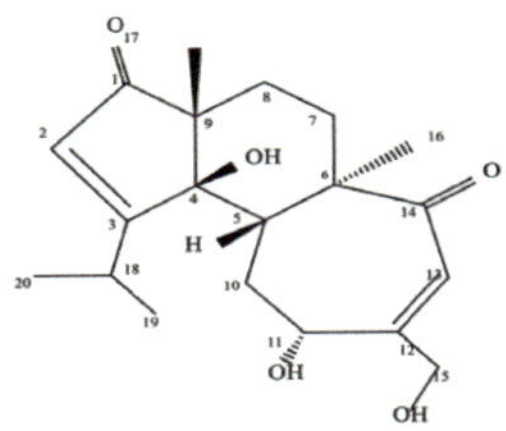
(24)

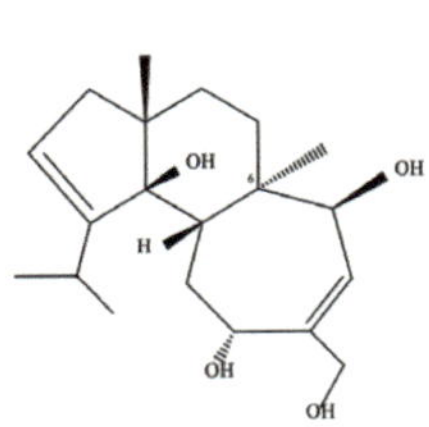
(25)

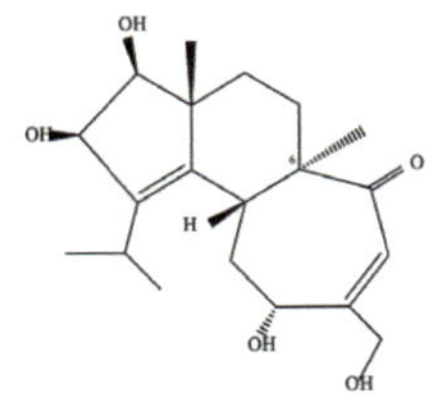
(26)

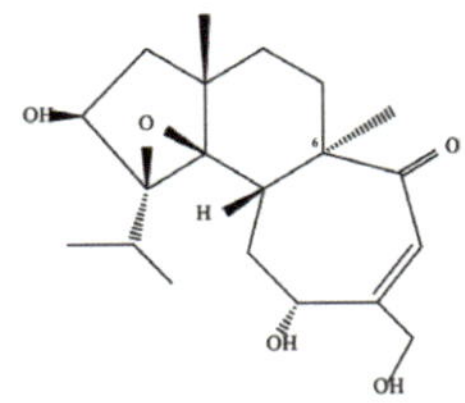
(27)

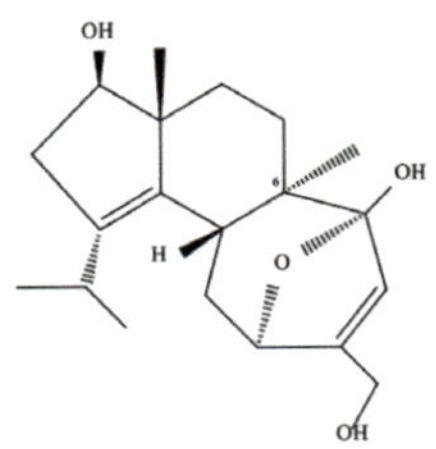
(28)

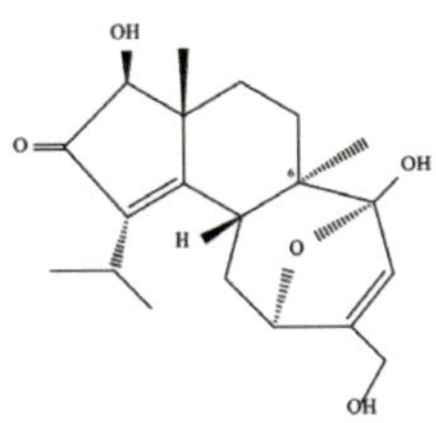
(29)

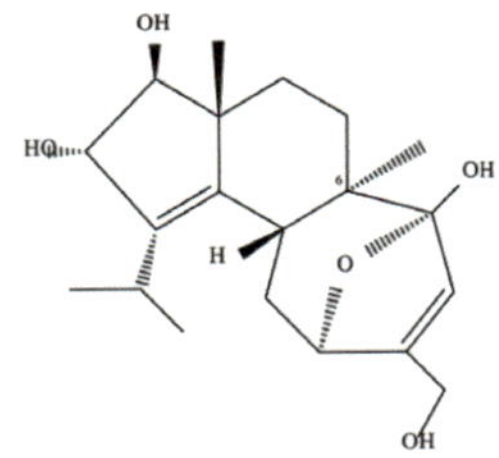
(30)

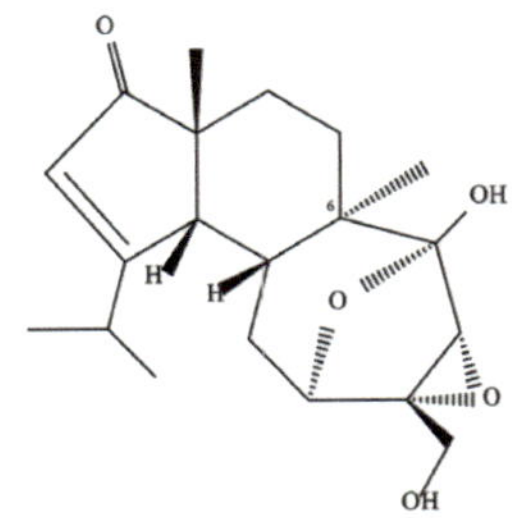
(31)

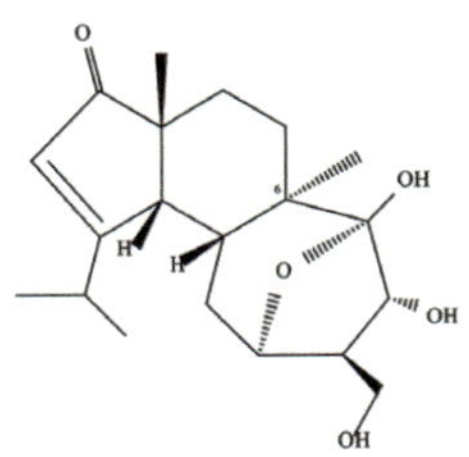
(32)

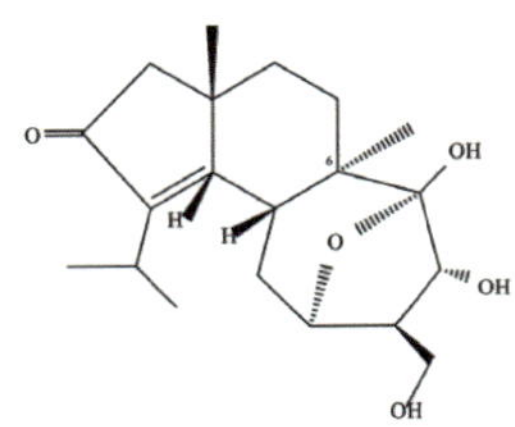
(33)

Figure 13. Structures of metabolites isolated from *C. africanus* [90].

in survival, maintenance, and regeneration of specific neuronal populations in the adult brain [95]. Therefore, the metabolites extracted from *Hericium* are important source of metabolites and source as remedy in the fight against neurodegenerative diseases such as Parkinson's, Alzheimer's, and Huntington's diseases, which are accompanied by decreased neurotrophic factor expression [102]. Two new potential metabolites have been isolated from *H. erinaceus* (strain STMA 06157B) and *H. flagellum* (strain CBS 103681) [96] (**Figure 14**).

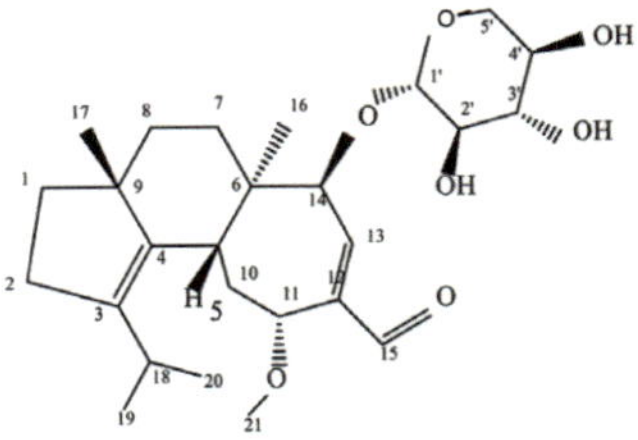
(34) Z1

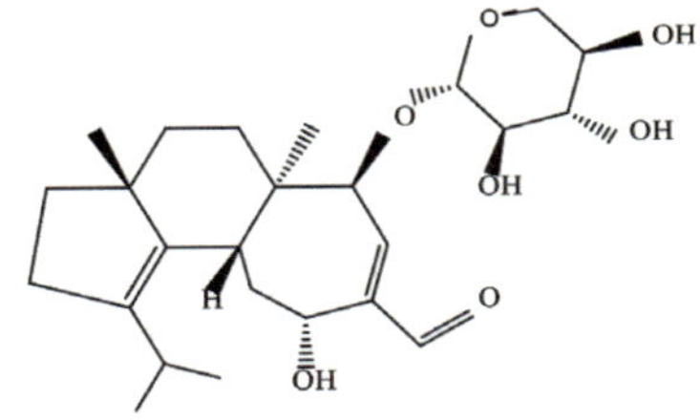
(35) Z2

Figure 14. Structural elucidation of metabolites isolated from *H. erinaceus* and *H. flagellum* [96].

Biological activity and ¹³C-NMR analysis: All of the metabolites isolated from *H. erinaceus* and *H. flagellum* exhibited strong neutrotrophin capacity [95, 96]. Metabolites were also studied through ¹³C-NMR; compound 34 exhibited the presence of five non-proton-bearing carbons, including three olefinic (δC 139.9, 136.6, 138.6) and two aliphatic carbons (δC 40.6, 49.2). Furthermore, five methylene groups with corresponding carbons between δC 28.4 and 38.5 ppm, a further oxygenated methylene group at δC 65.1, vicinal to two aliphatic methines at δC 40.4, and six methines at δC 69.6–105.3 ppm were observed. ¹³C shifts and correlations of the HSQC-DEPT spectrum showed high similarity to **35** which was a derivative of the cyathane diterpenoid **34**. The major difference between the two compound spectra was the missing methoxy group at C-11 in **35** (**Figure 13**). The detail of ¹³C-NMR data is described in **Table 2**.

3.3.2. *Indole diterpenes*

Indole diterpenes are the broad class of secondary metabolites with enormous structural and functional diversity. They mostly occur in filamentous fungal members having most abundance in *Penicillium, Aspergillus, Neotyphodium*, and *Claviceps* [97, 98]. This class of diterpenes is generally divided into two main groups, paxilline type and non-paxilline type [98], though it mainly consists of cyclic diterpenoid backbone in addition to an indole moiety.

3.3.2.1. *Diversity of indole diterpenes*

3.3.2.1.1. *Indole diterpenes from Aspergillus nidulans*

The marine fungi *A. nidulans* was reported to be the source of 19-hydroxypenitrem A (1) and 19-hydroxypenitrem E (2). The ¹³C-NMR spectrum of 19-hydroxypenitrem A ($C_{37}H_{44}ClNO_7$) provided 37 resonance states from 5 methyl, 8 methylene (with 2 sp_2 terminal), 1 sp_2 and 7 sp_3 methines (with 5 oxygenated), and 16 quaternary (with 5 oxygenated sp_3 and 9 sp_2) carbon atoms. In comparison, 19-hydroxypenitrem E ($C_{37}H_{45}NO_7$) lack chlorine atom but have one additional hydrogen atom [98] (**Figure 15**).

3.3.2.1.2. *Drechmeria sp.: a rich source of indole diterpenes*

An endophytic fungi *Drechmeria* sp. was found to be the reservoir of diverse indole diterpenes including drechmerin A (38), drechmerin B (39), drechmerin C (40), drechmerin D (41),

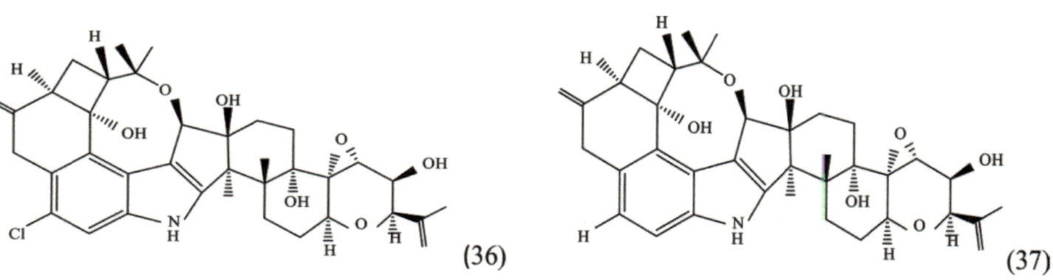

Figure 15. Indole diterpenes from *A. nidulans* [98].

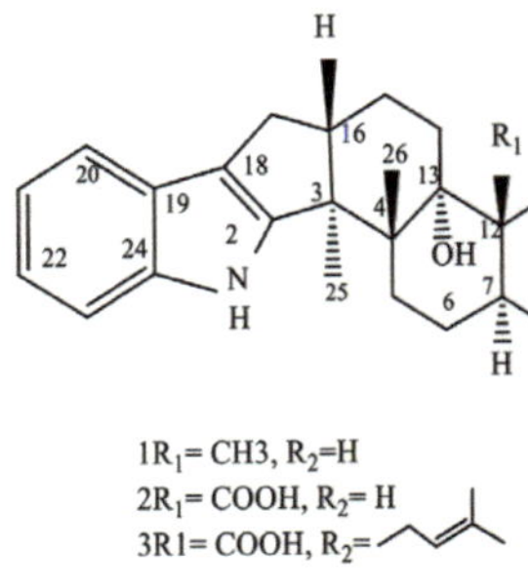

(38-40)

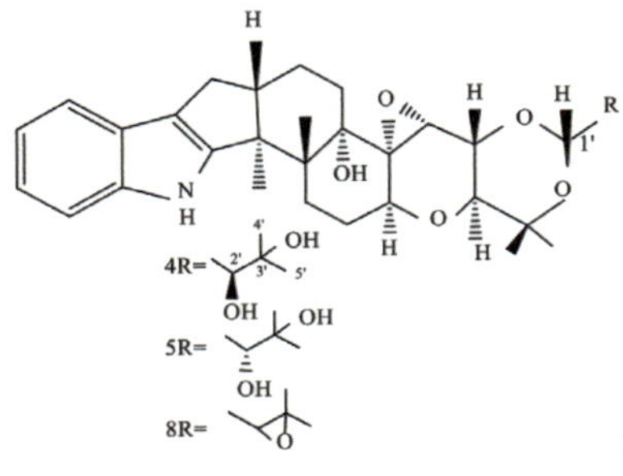

(41-42,45)

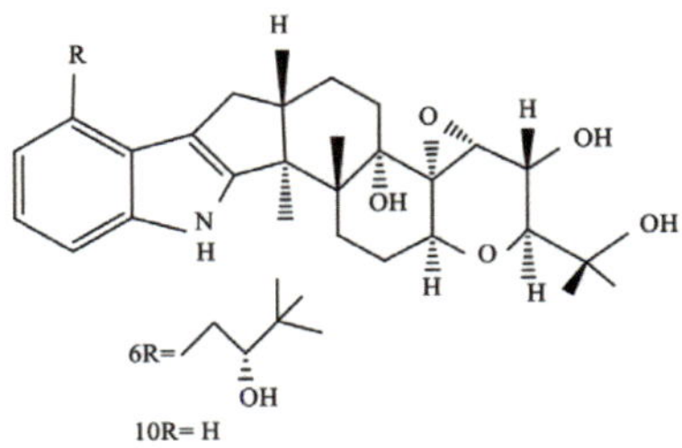

(43,47)

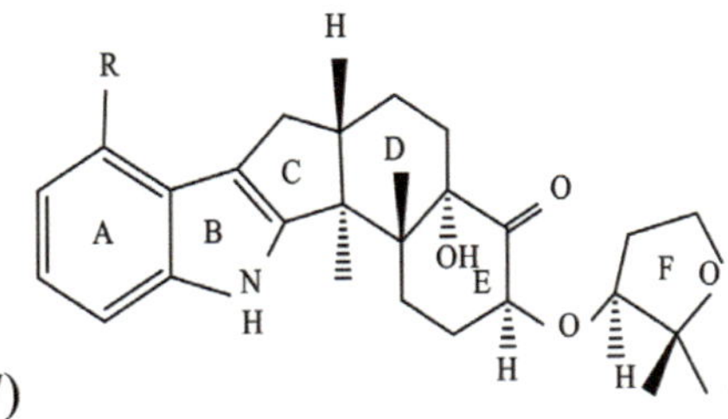

(44)

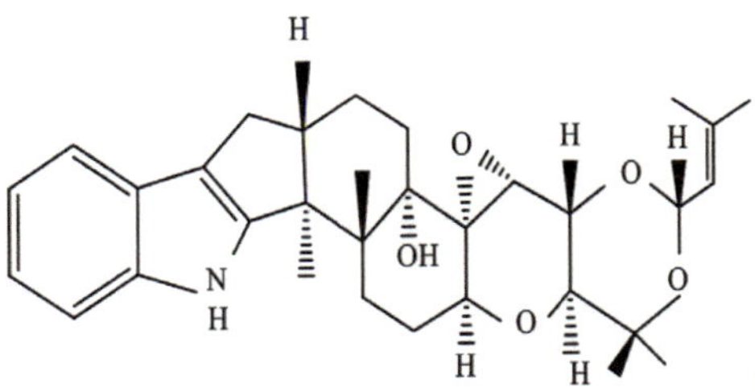

(45)

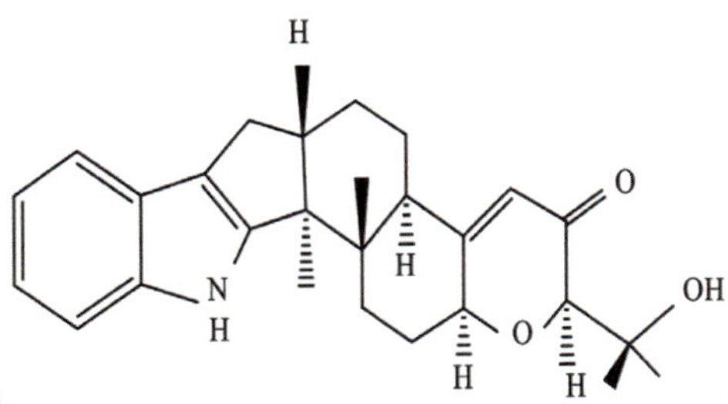

(48)

Figure 16. Indole diterpenes from (1–11) *Drechmeria* sp. [99].

drechmerin E (42), drechmerin F (43), drechmerin G (44), terpendole A (45), terpendole C (46), terpendole I (47), and dehydroxypaxilline (48) [99] (**Figure 16**).

The ^{13}C-NMR spectrum of drechmerin exhibited 28 carbon resonances, comprising 8 aromatic carbons, 4 oxygenated carbons, 6 methylene carbons, 2 methine carbons, 3 quaternary carbons, and 5 methyl carbons [100]. The detail of ^{13}C-NMR data is given in **Table 3**.

3.3.2.1.3. Indole diterpenes from marine A. flavus

The marine *Aspergillus flavus* had provided 4b-deoxy-β-aflatrem (1),9-isopentenyl paxilline (2), 6,8-di-*O*-methylcitreoisocoumarin (3), β-aflatrem (4), and paspaline (5). 4b-Deoxy-β-aflatrem

Carbon Number	Compound 36[g]	37[g]	38[c]	39[c]	40[c]	41[c]	42[c]	43[c]	44[c]
2	152.2	151.3	152.6	152.2	152.3	153.8	153.8	153.5	150.7
3	116.7	116.4	54.7	53.8	53.9	52.1	52.1	51.8	51.9
4	132.7	131.0	41.1	40.9	40.9	43.9	43.9	43.7	46.6
5	123.4	125.9	34.0	34.1	34.1	27.5	27.5	27.4	33.2
6	122.6	118.9	26.5	26.4	26.5	29.8	29.8	29.5	32.3
7	110.4	110.3	79.3	77.7	77.9	73.1	73.1	73.0	84.5
8	120.3	121.1	***	***	***	***	***	***	***
9	138.5	138.9	80.6	79.9	78.4	72.7	42.7	77.7	88.6
10	33.7	37.1	31.0	32.5	32.7	72.9	72.7	68.7	74.1
11	148.7	150.0	71.1	68.3	68.5	61.4	61.4	65.0	175.8
12	45.4	45.7	42.0	53.6	53.6	68.9	69.0	70.7	212.8
13	23.9	24.0	39.2	41.6	41.6	78.8	78.9	78.6	52.2
14	52.9	53.0	22.9	24.9	24.8	30.6	30.6	30.8	23.8
15	80.3	80.4	25.7	26.3	26.3	22.0	22.0	22.0	24.7
16	74.6	74.6	50.4	50.5	50.5	51.6	51.6	51.7	50.5
17			28.6	28.4	28.4	28.2	28.3	30.7	28.2
18	79.5	79.8	118.0	118.1	118.1	117.2	117.3	116.7	118.3
19	86.9	87.0	126.4	126.4	126.4	126.5	126.5	126.4	126.3
20	27.7	27.7	118.8	118.9	118.9	118.9	118.9	131.7	119.0
21	23.4	23.4	119.8	119.8	119.8	119.8	119.8	121.0	120.
22	75.9	76.0	120.8	120.9	120.9	120.8	120.8	121.0	121.2
23	64.8	64.8	112.8	112.8	112.8	112.7	112.7	110.8	112.8
24	60.0	60.0	142.2	142.2	142.2	141.9	141.9	142.0	142.2
25	64.7	64.7	15.0	15.0	15.0	16.6	15.6	16.5	15.0
26	73.0	73.0	20.4	17.0	17.1	19.1	19.1	19.0	16.4
27	***	***	14.1	178.2	178.2	***	***	***	***
28	69.7	69.8	73.0	72.8	77.7	76.2	75.1	73.7	16.4
29	29.0	29.0	25.5	26.4	23.7	28.7	28.8	27.5	84.0
30	27.9	28.0	25.9	25.5	22.1	17.2	17.1	25.2	27.8
1′	42.8	42.8			60.1	95.5	96.0	36.4	22.9
2′	53.8	53.9			123.7	77.5	77.6	78.4	
3′	106.8	105.3			136.3	77.9	77.8	79.0	
4′	19.3	19.3			18.2	22.6	22.4	22.0	

Carbon Number	Compound 36[g]	37[g]	38[c]	39[c]	40[c]	41[c]	42[c]	43[c]	44[c]
5′	29.0	29.0			26.1	21.3	21.7	20.6	
36	19.8	19.8							
37	142.5	142.5							
38	110.7	110.7							
39	19.9	19.8							
40	16.9	17.1							
Ref.	[98]	[98]	[99]	[99]	[99]	[99]	[99]	[99]	[99]

[a] ^{13}C-NMR at a 125 MHz.
[b] ^{13}C-NMR at 75 MHz in $CDCl_3$.
[c] ^{13}C-NMR at 150 MHz CD_3OD.
[g] ^{13}C-NMR at 125 MHz for ^{13}C, measured in DMSO-d6.

Table 3. ^{13}C-NMR data of indole diterpenoids.

($C_{32}H_{39}NO_3$) exhibits 14 degrees of unsaturation and consists of an indole chromophore and a carbonyl group. As per ^{13}C-NMR spectrum, the respective structure owns resemblance to β-aflatrem, except that a methine replaced an oxygenated quaternary carbon, thereby resulting in an isopentenylated indole diterpenoid. Moreover, 9-isopentenyl paxilline ($C_{32}H_{39}NO_4$) comprised hexacyclic indole diterpenoid skeleton [100] (**Figure 17**).

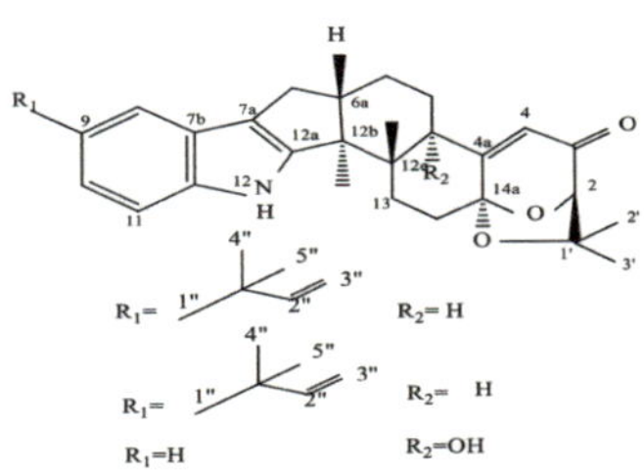
(48-50)

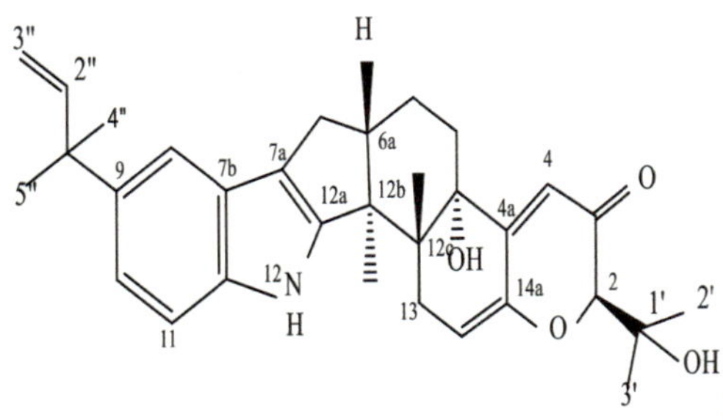
(51)

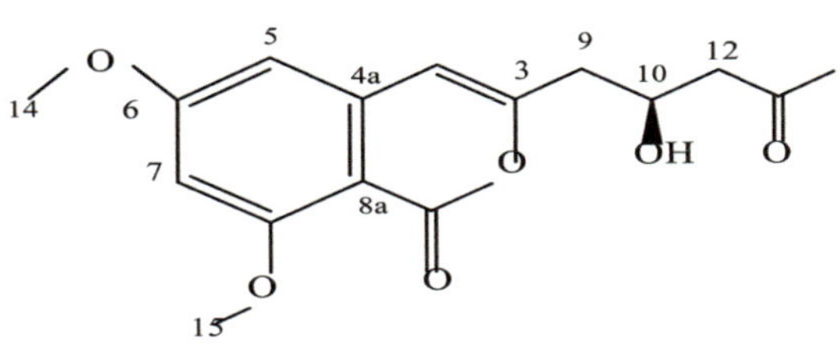
(52)

Figure 17. Skeleton by NMR indole diterpenes from *A. flavus* [100].

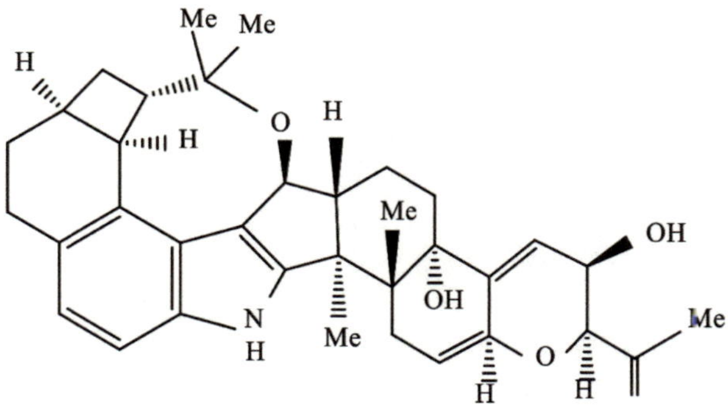

Figure 18. Structure and compound isolated from *P. crustosum* [102].

3.3.2.1.4. Penitrem D from Penicillium crustosum

Penitrem D ($C_{37}H_{45}NO_4$) was first isolated from *P. crustosum* in 1983. It is a complex structure with 9 rings, an indole core, and 11 stereocenters [101] (**Figure 18**).

3.3.2.1.5. Emindole SB from Emericella striata

The mycelium of *E. striata* was reported to naturally produce emindole SB ($C_{28}H_{39}NO$). In its structure an indole unit fused to a tricyclic carbon scaffold, and it presented six stereocenters, including vicinal quaternary centers on the western cyclohexyl ring [103] (**Figure 19**).

3.3.2.1.6. Paspaline obtained from Claviceps paspali

The ergot fungus *Claviceps paspali* was found to be the source of paspaline ($C_{28}H_{39}NO_2$) in 1966. The structure owes similarity with emindole SB but contains one more ring comparatively [104] (**Figure 20**).

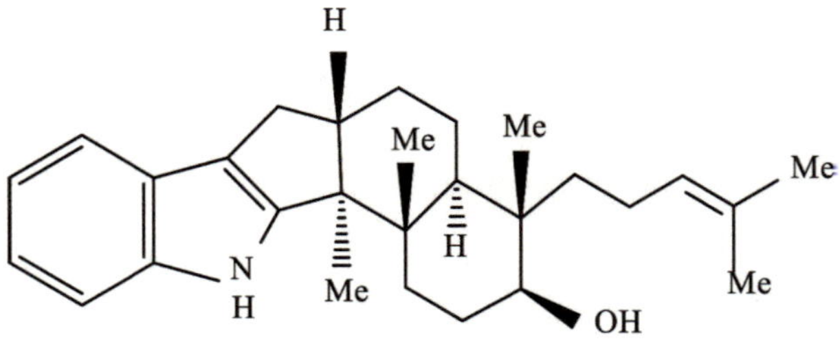

(54)

Figure 19. Structure and compound isolated from E. striata [103].

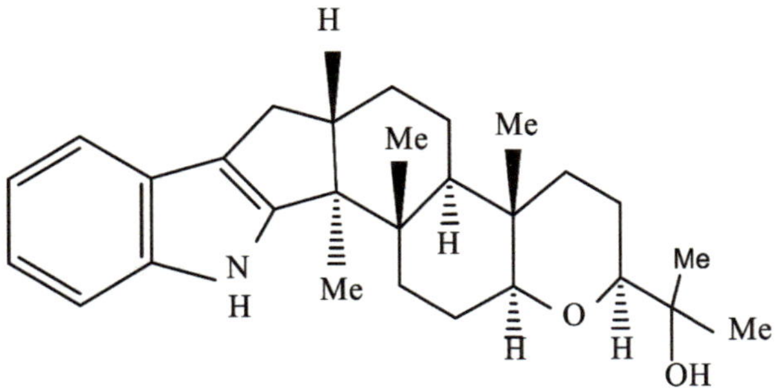

Figure 20. Structure and compound isolate from *Claviceps paspali* [104].

3.3.2.2. Biological activity

The indole diterpenes, famously called tremorgenic mycotoxins, put forward promising insecticidal potential via regulation of their glutamategated chloride ion channels [105], antibiotic activity [107, 108], antiproliferative against human breast cancer cells [109], and antifungal efficacy [110].

Author details

Fozia Farhat*, Arneeb Tariq, Annam Zikrea and Riffat Nasim Fatima

*Address all correspondence to: foziafarhat01@gmail.com

Department of Botany, Government College Women University, Faisalabad, Pakistan

References

[1] Dairi T. Studies on biosynthetic genes and enzymes of isoprenoids produced by actinomycetes. The Journal of Antibiotics. 2005;**58**:227-243

[2] Daum M, Herrmann S, Wilkinson B, Bechthold A. Genes and enzymes involved in bacterial isoprenoid biosynthesis. Current Opinion in Chemical Biology. 2009;**13**:180-188

[3] Fraga BM. Natural Sesquiterpenes. Natural Product Reports. 2011;**28**:1580-1610

[4] Xiao W, Li R, Huang S, Pu J, Sun H. Terpenoids from Schisandraceae family. Natural Product Reports. 2008;**25**:871-891

[5] Geris R, Simpson TJ. Meroterpenoids produced by fungi. Natural Product Reports. 2009;**26**:1063-1094

[6] Sun H, Li X, Meng L, Cui C, Gao S, Li C, et al. Asperolides A-C tetranorlabdane diterpenoids from marine algae- derived endophytic fungus *Aspergillus wentii* EN-48. Journal of Natural Products. 2012;**75**:148-152

[7] Liang CH, Syu JL, Mau JL. Antioxidant properties of solid-state fermented adlay and rice by *Phellinus linteus*. Food Chemistry. 2009;**116**:841-845

[8] Martins S, Mussatto SI, Martinez-Avila G, Montañez-Saenz J, Aguilar CN, Teixeira JA. Bioactive phenolic compounds: Production and extraction by solid-state fermentation. Biotechnology Advances. 2011;**29**:365-373

[9] Okamura-Matsui T, Izuta H, Tomoda T, Noda H, Fukuda S, Ohsugi M. Fermented soybean with thrombosis preventing activity using mushroom mycelia as microbial source. Food Science and Technology Research. 2003;**9**:227-230

[10] Zhang GP, Zhang F, Ru WM, Han JR. Solid-state fermentation of cornmeal with the ascomycete *Morchella esculenta* for degrading starch and upgrading nutritional value. World Journal of Microbiology and Biotechnology. 2010;**26**:15-20

[11] Zhang ZY, Lei ZF, Lu Y, Lu ZZ, Chen Y, Lei ZF. Chemical composition and bioactivity changes in stale rice after fermentation with *Cordyceps sinensis*. Journal of Bioscience and Bioengineering. 2008;**106**:188-193

[12] Rai M, Tidke G, Wasser SP. Therapeutic potential of mushrooms. Natural Product Radiance. 2005;**4**:246-257

[13] Bao L, Li YX, Wang QX, Han JJ, Yang XL, Li HR, et al. Nutritive and bioactive components in rice fermented with the edible mushroom *Pleurotus eryngii*. Mycology. 2013;**4**: 96-102

[14] Li YX, Bao L, Song B, Han JJ, Li HR, Zhao F, et al. A new benzoquinone and a new benzofuran from the edible mushroom *Neolentinus lepideus* and their inhibitory activity in NO production inhibition assay. Food Chemistry. 2013;**141**:1614-1618

[15] Wang YQ, Bao L, Yang XL, Li L, Li SF, Gao H, et al. Bioactive sesquiterpenoids from the solid culture of the edible mushroom *Flammulina velutipes* growing on cooked rice. Food Chemistry. 2012;**132**(3):1346-1353

[16] Joseph-Nathan Mulia P, Cuevas G, Quijano L. Journal of Natural Products. 2015;**78**: 2580-2587

[17] Tudznski B. Gibberellin biosynthesis in fungi: Genes, enzymes, evolution, and impact on biotechnology. Applied Microbiology and Biotechnology. 2005;**66**:597-611

[18] Kirby J, Keasling JD. Biosynthesis of plant isoprenoids: Perspectives for microbial engineering. Annual Review of Plant Biology. 2009;**60**:335-355

[19] Christianson DW. Chemical Reviews. 2006;**106**:3412-3442

[20] De Boer AH, de Vries-van Leeuwen IJ. Fusicoccanes: Diterpenes with surprising biological functions. Trends in Plant Science. 2012;**17**(6):360-368

[21] Toyomasu T et al. Fusicoccins are biosynthesized by an unusual chimera diterpene synthase in fungi. Proceedings of the National Academy of Sciences of the United States of America. 2007;**104**(9):3084-3088

[22] Peters RJ. Two rings in them all: The labdane-related diterpenoids. Natural Product Reports. 2010;**27**:1521-1530

[23] Wilderman PR, Peters RJ. A single residue switch converts abietadiene synthase into a pimaradiene specific cyclase. Journal of the American Chemical Society. 2007;**129**(51): 15736-15737

[24] Xu M, Wilderman PR, Peters RJ. Following evolution's lead to a single residue switch for diterpene synthase product outcome. Proceedings of the National Academy of Sciences of the United States of America. 2007;**104**(18):7397-7401

[25] Li R, Morris-Natschke SL, Lee KH. Clerodane diterpenes: Sources, structures and biological activities. Natural Product Reports. 2016;**33**:1166-1226

[26] Monteiro AF, Batista JM, Machado MA, Severino RP, Bianah EW, Bolzani VS, et al. Diterpenoids of terrestrial origin. Journal of Natural Products. 2015;**78**:1451-1455

[27] Wu H, Wang S, Xu Z, Sun S, Liu H, Wang J, et al. Two new clerodane diterpene from *Tinospora sagitatta*. Molecules. 2015;**20**:839-845

[28] Annan K, Ekuadzi E, Asare C, Sarpong K, Pistorius D, Oberer L, et al. Phytochemistry Letters. Claerodane diterpene from *Polyalthia logifolia* (Sonn). 2015;**11**:28-31

[29] Nohra YA, Perrichot V, Jeanneau L, LePolles L, Azar D. Chemical characterization and botanical origin of French ambers. Journal of Natural Products. 2015;**78**:1284-1293

[30] Vasconcelos DHP, Barbosa FG, Oliveira NDCF, Lima MAS, Freire FDCO, et al. New clerodane diterpenes from fungal biotransformation of the 3,12-dioxo-15,16-epoxy-4-hydroxycleroda-13(16),14-diene. MOJ Bioorganic and Organic Chemistry. 2017;**1**(6):00036. DOI: 10.15406/mojboc.2017.01.00036

[31] Monte FJQ, Dantas EMG, Braz-Filho R. New diterpenoids from croton argyrophylloides. Phytochemistry. 1988;**27**(10):3209-3212

[32] Mafezoli J, Oliveira MC, Paiva JR, Sousa AH, Lima MA, et al. Stereo and regioselective microbial reduction of the clerodane diterpene 3,12-dioxo-15,16-epoxy-4-hydroxycleroda-13(16),14-diene. Natural Product Communications. 2014;**9**(6):759-762

[33] Gebbinck K, Klein EA, Jansen BJM, de Groot A. Phytochemistry. 2002;**61**:737-770

[34] Demetzos C, Konstantinos SD. Labdane type diterpenes: Chemistry and biological activity. Studies in Natural Products Chemistry. 2001;**25**:235-292

[35] de Vargas FS, de Almeida PDO, Aranha ESP, de Boleti AP, Newton P, de Vasconcelios MC, et al. Biological activities and cytotoxicity of diterpenes from *Copaifera spp.* Oleorasins. Molecules. 2015;**20**:6194-6210

[36] Leandro LF, Cardoso MJO, Silva SDC, Souza MGM, Veneziani RCS, Ambrosio SR, et al. Antibacterial activity of pinus elloitti and its major compounds, dehydroabietic acid against multidrug resistant strain. Medical Microbiology. 2014;**63**:1649

[37] da Silva SD, Mendes de Souza MG, Oliveira Cardoso MJ, da Silva Moraes T, Ambrósio SR, Sola Veneziani RC, et al. Antibacterial activity of pinus elliotti against anaerobic bacteria. Anaerobe. 2014;**30**:146

[38] Manner S, Vahermo M, Skogman ME, Krogerus S, Vuorela PM, Yli-Kauhaluoma J, et al. New derivatives of dehydroabietic acid target planktonic and biofilm bacteria in staphylococcus aureus and effectively disrupt bacterial membrane integrity. The European Journal of Medicinal Chemistry. 2015;**102**:68

[39] Burgstahler AW, Marx JN. Digenesis of Tricyclic diterpenes. The Journal of Organic Chemistry. 1969;**34**:1562-1566

[40] Gonzalez JN, Perez-Guaita D, Correa-Royero J, Zapata B, Agudelo L, Mesa-Arango M, et al. Synthesis and biological evolution of of novel Dehydoabietic acid dericative conjugated with acyl thiourea peptide moiety as antitumor agent. European Journal of Medicinal Chemistry. 2010;**45**:811-816

[41] Gonzalez MA. Aromatic abietane diterpenoids: Their biological activity and synthesis. Natural Product Reports. 2012:1-17

[42] Helfenstein A, Vahermo M, Nawrot DA, Demirci F, Gökalp I, Krogerus S, et al. Antibacterial profiling of abietane-type diterpenoids. Bioorganic and Medicinal Chemistry. 2017;**25**:132-137

[43] Fallarero A, Skogman M, Kujala J, Rajaratnam M, Moreira VM, Yli-Kauhaluoma J, et al. The International Journal of Molecular Sciences. 2013;**14**:12054-12072

[44] Jang HJ, Yang KS. Archives of Pharmacal Research. 2011;**34**:913-917

[45] Kawada T, Hirai S, Goto T, Kuroyanagi K, Kim Y, Ohyama K, et al. BioFactors. 2009;**35**: 442-448

[46] Muhammad I, Mossa JS, El-Feraly FS. Phytotherapy Research. 1992;**6**:261-264

[47] Chang S, Chen P, Wang S, Wu H. The Journal of Medical Entomology. 2001;**38**:455-457

[48] Ulubelen A, Birman H, Oksuz S, Topcu G, Kolak U, Barla A, et al. Planta Medica. 2002; **68**:818-821

[49] Ono M, Yamamoto M, Masuoka C, Ito Y, Yamashita M, Nohara T. Journal of Natural Products. 1999;**62**:1532-1537

[50] Areche C, Theoduloz C, Yanez T, Souza-Brito ARM, Barbastefano V, de Paula D, et al. Journal of Pharmacy and Pharmacology. 2008;**60**:245-251

[51] Iwamoto M, Minami T, Tokuda H, Ohtsu H, Tanaka R. Planta Medica, 2003;**69**:69-72

[52] Bispo de Jesus M, Zambuzzi WF, Ruela de Sousa RR, Areche C, Santos de Souza AC, Aoyama H, et al. Biochimie. 2008;**90**:843-854

[53] Gough LJ. Tetrahedron Letters. 1968;**3**:295-298

[54] Carman RM, Deeth HC. The Australian Journal of Chemistry. 1967;**20**:2789-2793

[55] Zhang GJ, Li YH, Jiang JD, Yu SS, Qu J, Ma SG, et al. Tetrahedron. 2013;**69**:1017-1023

[56] Wang P, Deng G, Yuan W, Su Z. Bioorganic and Medicinal Chemistry Letters. 2013;**23**: 6682-6687

[57] Aruoma OI, Spencer JPE, Rossi R, Aeschbach R, Khan A, Mahmood N, et al. Food and Chemical Toxicology. 1996;**34**:449-456

[58] Horiuchi K, Shiota S, Kuroda T, Hatano T, Yoshida T, Tsuchiya T. Biological and Pharmaceutical Bulletin. 2007;**30**:287-290

[59] Xing V, Wu H, Wang X, Huang Y, Li Q, Li C, et al. Journal of Chemotherapy. 2008;**20**: 238-245

[60] Kobayashi K, Nishino C, Tomita H, Fukushima M. Phytochemistry. 1987;**26**:3175-3179

[61] Ayer WA, Taube H. Metabolites of *Cyathus helenae*. Cyathin A3 and allocyathin B3, members of a new group of diterpenoids. Tetrahedron Letters. 1972;**19**:1917-1920

[62] Wright DL, Whitehead CR. Recent progress on the synthesis of cyathane type diterpenes. Organic Preparations and Procedures International. 2000;**32**:307-331

[63] Krzyczkowshi W, Malinowska E, Herold F. The structure, medicinal properties and biosynthesis of cyathane diterpenoids. Biotechnologia. 2008;**1**:146-147

[64] Kamo T, Imura Y, Hagio T, Makabe H, Shibata H, Hirota M. Antiinflammatory cyathane diterpenoids from *Sarcodon scabrosus*. Bioscience, Biotechnology, and Biochemistry. 2004; **68**:1362-1365

[65] Hirota M, Morimura K, Shibata H. Anti-inflammatory compounds from the bitter mushroom, *Sarcodon scabrosus*. Bioscience, Biotechnology, and Biochemistry. 2002;**66**:179-184

[66] Shi XW, Liu L, Gao JM, Zhang AL. Cyathane diterpenes from Chinese mushroom *Sarcodon scabrosus* and their neurite outgrowth-promoting activity. European Journal of Medicinal Chemistry. 2011;**46**:3112-3117

[67] Ma BJ, Liu JK. A new bitter diterpenoid from *Sarcodon scabrosus*. Journal of Basic Microbiology. 2005;**45**:328-330

[68] Curini M, Maltese F, Marcotullio MC, Menghini L, Pagiotti R, Rosati O, et al. Glaucopines A and B, new cyathane diterpenes from the fruiting bodies of *Sarcodon glaucopus*. Planta Medica. 2005;**71**:194-196

[69] Marcotullio MC, Pagiotti R, Campagna V, Maltese F, Fardella G, Altinier G, et al. Glaucopine C, a new diterpene from the fruiting bodies of *Sarcodon glaucopus*. Natural Product Research. 2006;**20**:917-921

[70] Marcotullio MC, Pagiotti R, Maltese F, Oball-Mond Mwankie GN, Hoshino T, Obara Y, et al. Cyathane diterpenes from *Sarcodon cyrneus* and evaluation of their activities of neuritegenesis and nerve growth factor production. Bioorganic and Medicinal Chemistry. 2007;**15**:2878-2882

[71] Marcotullio MC, Pagiott R, Maltese F, Obara Y, Hoshino T, Nakahata N, et al. Neurite outgrowth activity of cyathane diterpenes from *Sarcodon cyrneus*, cyrneines A and B. Planta Medica. 2006;**72**:819-823

[72] Ayer WA, Browne LM, Mercer JR, Taylor DR, Ward DE. Metabolites of bird's nest fungi. Part 8. Some minor metabolites of *Cyathus helenae* and some correlations among the cyathins. Canadian Journal of Chemistry. 1978;**56**:717-721

[73] Ayer WA, Yoshida T, van Schie DMJ. Metabolites of bird's nest fungi. Part 9. Diterpenoid metabolites of *Cyathus africanus Brodie*. The Canadian Journal of Chemistry. 1978;**56**: 2113-2120

[74] Ayer WA, Lee SP. Metabolites of bird's nest fungi. Part 11. Diterpenoid metabolites of *Cyathus earlei*. Canadian Journal of Chemistry. 1979;**57**:3332-3337

[75] Hecht HJ. Striatin A, B, and C, novel diterpenoid antibiotics from *Cyathus striatus*, X-ray crystal structure of striatin A. Journal of the Chemical Society, Chemical Communications. 1978;**15**:665-666

[76] Shiono Y, Hiramatsu F, Murayama T, Koseki T, Funakoshi T. Two cyathane-type diterpenoids from the liquid culture of *Strobilurus tenacellus*. Chemistry and Biodiversity. 2008;**5**:1811-1816

[77] Kawagishi H, Simada A, Hosokawa S, Mori H, Sakamoto H, Ishiguro Y, et al. Erinacines E, F, and G, stimulators of nerve growth factor (NGF)-synthesis, from the mycelia of *Hericium erinaceum*. Tetrahedron Letters. 1996;**37**:7399-7402

[78] Kawagishi H, Simada A, Shizuki K, Mori H, Sakamoto H, Furukawa S. Erinacine D, a stimulator of NGF-synthesis, from the mycelia of *Hericium erinaceum*. Heterocyclic Communications. 1996;**2**:51-54

[79] Kawagishi H, Shimada A, Shirai R, Okamoto K, Ojima F, Sakamoto H, et al. Erinacines A, B and C, strong stimulators of nerve growth factor (NGF)-synthesis, from the mycelia of *Hericium erinaceum*. Tetrahedron Letters. 1994;**35**:1569-1572

[80] Kawagishi H, Masui A, Tokuyamab S, Nakamura T. Erinacines J and K from the mycelia of *Hericium erinaceum*. Tetrahedron. 2006;**62**:8463-8466

[81] Lee EW, Shizuki K, Hosokawa S, Suzuki M, Suganuma H, Inakuma T, et al. Two novel diterpenoids, erinacines H and I from the mycelia of *Hericium erinaceum*. Bioscience, Biotechnology, and Biochemistry. 2000;**64**:2402-2405

[82] Kenmoku H, Sassa T, Kato N. Isolation of erinacine P, a new parental metabolite of cyathane-xylosides, from *Hericium erinaceum* and its biomimetic conversion into erinacines A and B. Tetrahedron Letters. 2000;**41**:4389-4393

[83] Kenmoku H, Shimai T, Toyomasu T, Kato N, Sassa T. Erinacine Q, a new erinacine from *Hericium erinaceum*, and its biosynthetic route to erinacine C in the basidiomycete. Bioscience, Biotechnology, and Biochemistry. 2002;**66**:571-575

[84] Kenmoku H, Kato N, Shimada M, Omoto M, Mori A, Mituhashi W, et al. Isolation of (−)-cyatha-3,12-diene, a common biosynthetic intermediate of cyathane diterpenoids, from an erinacine-producing basidiomycete, *Hericium erinaceum*. Tetrahedron Letters. 2001;**42**:7439-7422

[85] Wang B, Han J, Xu W, Chen Y, Liu H. Production of bioactive cyathane diterpenes by a bird's nest fungus *Cyathus gansuensis* growing on cooked rice. Food Chemistry. 2014;**152**(2014):169-176

[86] Xu J, Jin DQ, Song H, Guo YQ, He YS. Lathyrane diterpenes from *Euphorbia prolifera* and their inhibitory activities on LPS-induced NO production. Fitoterapia. 2012;**83**:1205-1209

[87] He L, Han J, Li B, Huang L, Ma K, Chen Q, et al. Identification of a new cyathane diterpene that induces mitochondrial and autophagy-dependent apoptosis and shows a potent in vivo anticolorectal cancer activity. European Journal of Medicinal Chemistry. 2016;**111**:183-192

[88] Han JJ, Zhang L, Xu JK, Bao L, Zhao F, Chen YH, et al. Three new cyathane diterpenoids from the medicinal fungus *Cyathus africanus*. Journal of Asian Natural Products Research. 2013;**17**:541-549

[89] Han J, Chen YH, Bao L, Yang XL, Liu D, et al. Anti-inflammatory and cytotoxic cyathane diterpenoids from the medicinal fungus *Cyathus africanus*. Fitoterapia. 2015;**84**:22-31

[90] Wei J, Cheng Y, Guo WH, Wang DC, Zhang Q, Li D, et al. Molecular diversity and potential anti-neuroinflammatory activities of cyathane diterpenoids from the basidiomycete *Cyathus africanus*. Scientific Reports. 2017;**21**:8883. DOI: 10.1038/s41598-017-09118-z

[91] Han J, Chen Y, Bao L, Yang X, Liu D, Li S, et al. Anti-inflammatory and cytotoxic cyathane diterpenoids from the medicinal fungus *Cyathus africanus*. Fitoterapia. 2013;**84**:22-31

[92] Block ML, Hong JS. Microglia and inflammation-mediated neurodegeneration: Multiple triggers with a common mechanism. Progress in Neurobiology. 2005;**76**:77-98. DOI: 10.1016/j.pneurobio.2005.06.004

[93] Giovannini MG et al. Experimental brain inflammation and neurodegeneration as model of Alzheimer's disease: Protective effects of selective COX-2 inhibitors. International Journal of Immunopathology and Pharmacology. 2003;**16**:31-40

[94] Young BM. Antiplasmodial activity: The first proof of inhibition of hemecrystallization by marine isonitriles. European Journal of Medicinal Chemistry. 2015;**93**(2015):373-380

[95] Allen SJ, Watson JJ, Shoemark DK, Barua NU, Patel NK. GDNF, NGF and BDNF as therapeutic options for neurodegeneration. Pharmacology and Therapeutics. 2013;**138**:155-175

[96] Rupcic Z, Rascher M, Kanaki S, Köster RW, Stadler M, Wittstein K. Two new cyathane diterpenoids from mycelial cultures of the medicinal mushroom *Hericium erinaceus* and the rare species, *Hericium flagellum*. The International Journal of Molecular Sciences. 2018;**19**:740. DOI: 10.3390/ijms19030740

[97] Saikia S, Nicholson MJ, Young C, Parker EJ, Scott B. The genetic basis for indole-diterpene chemical diversity in filamentous fungi. Mycological Research. 2008;**112**:184-199

[98] Zhang P, Li XM, Li X, Wanga BG. New indole-diterpenoids from the algal-associated fungus *Aspergillus nidulans*. Phytochemistry Letters. 2015;**12**:182-185

[99] Zhao JC, Wang YL, Zhang TY, Chen ZJ, Yang TM, Wu YY, et al. Indole diterpenoids from the endophytic fungus *Drechmeria* sp. as natural antimicrobial agents. Phytochemistry. 2018;**148**:21-28

[100] Sun K, Li Y, Guo L, Wang Y, Liu P, Zhu W. Indole diterpenoids and isocoumarin from the fungus, *Aspergillus flavus*, isolated from the prawn, *Penaeus vannamei*. Marine Drugs. 2014;**12**:3970-3981

[101] Sallam AA, Houssen WE, Gissendanner CR, Orabi KY, Foudah AI, El-Sayed KA. Bioguided discovery and pharmacophore modeling of the mycotoxic indole diterpene alkaloids penitrems as breast cancer proliferation, migration, and invasion inhibitors. Medicinal Chemistry Communications. 2013;**4**(10):1360-1369

[102] Calvo AM, Cary JW. Association of fungal secondary metabolism and sclerotial biology. Frontiers in Microbiology. 2015;**6**:62

[103] Tomoda H, Tabata N, Yang DJ, Takayanagi H, Omura S. Terpendoles, novel ACAT inhibitors produced by *Albophoma yamanashiensis*. III. Production, isolation and structure elucidation of new components. The Journal of Antibiotics. 1995;**48**(8):793-804

[104] Jesus AE, Gorst-Allman CP, Steyn PS, Heerden FR, Vleggaar R, Wessels PL, et al. Tremorgenic mycotoxins from *Penicillium crustosum*: Biosynthesis of penitrem A. Journal of the Chemical Society, Perkin Transactions. 1983;**1**:1863-1868

[105] George DT, Kuenstner EJ, Pronin SV. A concise approach to Paxilline indole diterpenes. Journal of the American Chemical Society. 2015;**137**(49):15410-15413

[106] Smith III AB, Visnick M. An expedient synthesis of substituted indoles. Tetrahedron Letters. 1985;**26**(32):3757-3760

[107] Smith MM, Warren VA, Thomas BS, Brochu RM, Ertel EA, Rohrer S, et al. Nodulisporic acid opens insect glutamate-gated chloride channels: Identification of a new high affinity modulator. Biochemistry. 2000;**39**:5543-5554

[108] Dowd PF, Cole RJ, Vesonder RF. Toxicity of selected tremorgenic mycotoxins and related compounds to *Spodoptera frugiperda* and *Heliothis zea*. The Journal of Antibiotics. 1988;**41**:1868-1872

[109] Ondeyka JG, Helms GL, Hensens OD, Goetz MA, Zink DL, Tsipouras A, et al. Nodulisporic acid A, a novel and potent insecticide from a *Nodulisporium* sp. Isolation, structure determination and chemical transformations. Journal of the American Chemical Society. 1997;**119**:8809-8816

[110] Shen JW, Ruan Y, Ma BJ. Diterpenoids of macromycetes. Journal of Basic Microbiology. 2009;**49**(3):242-255